化学英語101

リスニングとスピーキングで効率的に学ぶ

國安 均 著 ／ 英語監修 ジェフリー・M・ストライカー

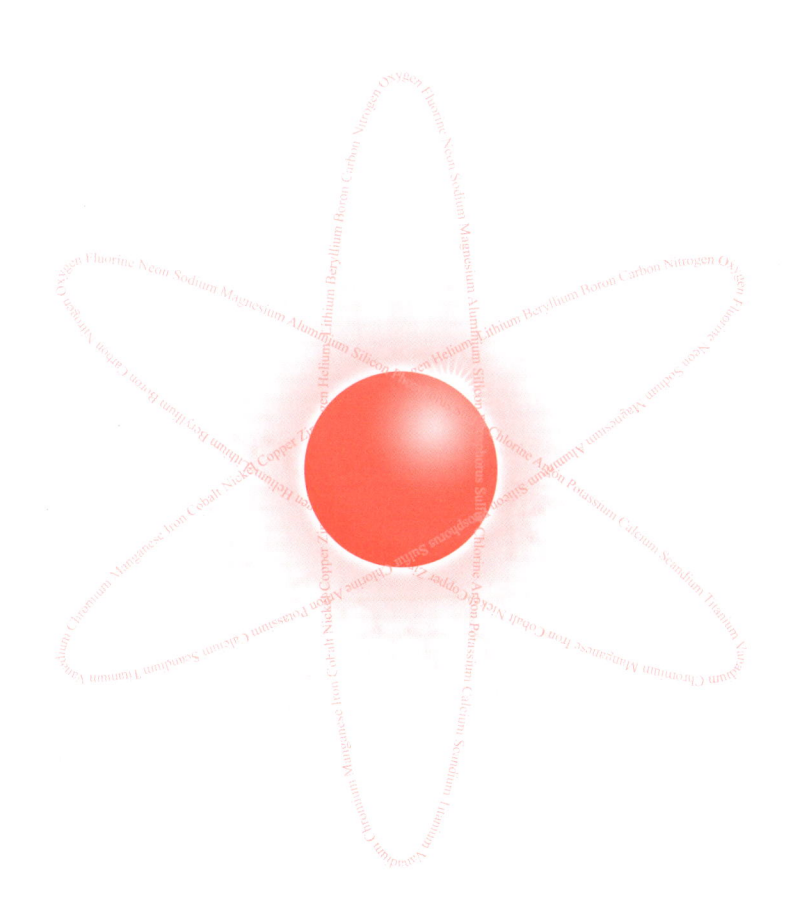

化学同人

PREFACE

▶ 本書出版の目的
　本書は「化学」を題材にして,「読む・書く・聞く・話す」すべての英語力を飛躍的に向上させることを目論んで執筆した学習書である．書名の「101」は，大学の入門用の授業につけられるコード番号である．つまり本書は，学部学生に化学英語の基礎を総合的に身につけてもらうための書物である．しかしながら，大学院生やすでに化学の仕事に携わっておられる方々にも十分に満足していただける内容に仕上がっているものと確信する．

▶ 化学と英語
　自然科学分野で立身するなら英語との付き合いを避けて通ることは難しい．化学も例にもれずである．論文は英語で書かれたものがほとんどであるし，大学院生であれば，留学生と接したり，英語の講演を聞いたりする機会もあるだろう．社会人であれば，海外出張・海外滞在することも今では珍しいことではない．著者も，学部3年生のときに化学英語に関連した授業を受講して以来，現在に至るまで，いろいろな形で英語と接し続けている．院生のときには，「化学英語」に関連した書物を何冊か購入し，英語力を高めようと努めた．しかし，そういった教材のほとんどは，「科学」あるいは「化学」を対象にしているとはいえ，文法を中心テーマとした教材が多く，化学に関する英語を聞き取ったり，声に出したりといった英語の基本的な能力の向上が望めるものではない．断片的あるいは，余計な説明（ノイズ）が多い書物がほとんどで，化学英語のコアーとなるべき学習方法についてわかりやすく述べられた書物は見当たらない．また，最低限必要な数の重要化学用語を網羅しているものもない．結局，英語で書かれた論文や教科書の中の単語を，辞書や用語集でコツコツ調べて意味や用法を覚えていくという効率の悪い方法が，化学英語学習の王道であるとされているように思う．

　自らの恥をさらすようではあるが，著者自身，博士課程修了時の英会話力は初心者レベルであり，自分の専門分野の化学について語ることさえ，英語ではおぼつかなかった．すでに英語で論文を数報執筆していたにもかかわらずである．論文を読み書きする力は高められても，コミュニケーション力を含めた英語力向上には，学生時代の学習方法はほとんど役立っていなかったのである．卒業後，英会話をはじめ，米国へ留学し，学生に化学英語を教える立場になって，自分の英語学習歴を回顧してみると，化学英語学習を上手く活用することにより，リスニングやスピーキングの力も含めた総合的な英語力を，もっと効率よく向上させるすべがあったのではないかと思う．

今でも英語に精通しているとは到底言いがたい著者ではあるが，今までの経験を踏まえ，これから化学英語を学ぼうとしている人や，今まさに独学で苦労している人々に，化学英語を通じて英語力を飛躍的にアップさせるすべを提供すべく本書を執筆した次第である．本書を活用することで，一人でも多くの方々が，英語に対して新たな関心を抱き，よりいっそう自らの英語力を鍛錬する意欲がかき立てられるとすれば本望である．

　なお本書は，平成 15 年度から大阪大学工学部応用自然科学科応用化学コース 2 年生後期に実施されている，基礎化学英語学習のための授業「ゼミナール I」の補助テキストとして製作したものに大幅に筆を加え，編集し直したものである．

▶ 謝　辞

　本書を執筆するにあたっては，益山新樹 大阪大学助教授（現 大阪工業大学教授）を中心とする，大阪大学工学部応用自然科学科応用化学コースの「ゼミナール改革ワーキンググループ」の多くの先生方に貴重なご助言をいただいた．また，本書の内容については，The University of Alberta 化学科の Jeffrey M. Stryker 教授と森田将基博士に十分に吟味していただき，たいへん貴重なたくさんのコメントをいただいた．CD*の音声は，Sukanda Tianniam さんと洪愛薇さんに吹き込んでいただいた．また，出版にあたっては，後藤 南氏，柞井文子氏，加藤貴広氏をはじめ，化学同人の方々にいろいろご尽力いただいた．これらの方々に，ここに改めて感謝の意を表する次第である．

2007 年 9 月
國安　均

＊編集部注：CD で提供していた音声ファイルは化学同人 HP からダウンロードできるようになりました．

本書を授業用テキストとして利用する場合

　本書は自学自習できるよう構成されているが，学部学生の授業用テキストあるいは補助テキストとしても使える．その場合，Chapter 1 と Chapter 2 は，自習用としてもよい．Chapter 3 および Chapter 4 は，大学の 1 セメスターの通常の授業数を考慮して，それぞれ 15 のレッスンから成っている．この部分を毎週の課題とするなどすれば，効果的に学習できる．

CONTENTS

PREFACE …………………………………………………………………………… ii
本書の特長と効果的な学習法 ……………………………………………………… vi

Chapter 1　発音の基本から始めよう　　#1-24

1.1　音の構成から見た日本語と英語 ……………………………………………… 1
1.2　発音の基本 ……………………………………………………………………… 2
1.3　母音の発音 ……………………………………………………………………… 2
1.4　子音の発音 ……………………………………………………………………… 7
1.5　口の形から見た発音のまとめ ………………………………………………… 17

Chapter 2　化学の基本単語を覚えよう　　#25-51

2.1　化学で重要な接頭語と接尾語 ………………………………………………… 19
　　2.1.1　接頭語(prefix)　19
　　2.1.2　接尾語(suffix)　21
　　2.1.3　乗数　22
2.2　元素の名称 ……………………………………………………………………… 22
2.3　化合物の名称 …………………………………………………………………… 24
2.4　人名反応など …………………………………………………………………… 31
2.5　その他の反応名など …………………………………………………………… 33
2.6　化学にかかわる者として最低限知っておきたい英単語 …………………… 34
　　2.6.1　学生・研究生活関連の単語　34
　　2.6.2　学年・職位・学位などの単語　35
　　2.6.3　教科名などの単語　35
　　2.6.4　実験関連の単語　36

Chapter 3　化学に頻出する基本構文を覚えよう　　#52-66

3.1　英語の文章を声にだして読むときの注意 …………………………………… 37
3.2　実践トレーニング 構文150 (Lesson 1〜15) ………………………………… 38

Chapter 4　化学英語の文章に親しもう　#67-98

4.1　リスニング，スピーキングを用いた学習方法 …………………………………… 79
4.2　実践トレーニング 教科書の英文 15（Lesson 1〜15）……………………… 81

Chapter 5　化学の論文や記事を読みこなそう

5.1　英語長文の読解法 …………………………………………………………………… 117
5.2　実践トレーニング 英文和訳 7（Lesson 1〜7）……………………………… 118

Chapter 6　化学を英文で書き表そう

6.1　英作文を書くときの心得 …………………………………………………………… 123
6.2　化学反応を英語で書く ……………………………………………………………… 126
　　6.2.1　化学反応の記述法の基本　126
　　6.2.2　さまざまな表現をマスターしよう　127
6.3　辞書を使って表現に注意して書く ………………………………………………… 129
6.4　実践トレーニング 英作文 100（Lesson 1〜10）…………………………… 131

Chapter 7　大学院入試問題に挑戦しよう

7.1　大学院入試問題の傾向と対策 ……………………………………………………… 143
7.2　実践トレーニング 入試問題 7（Lesson 1〜7）……………………………… 144

　実践トレーニングの解答例 ……………………………………………………… 155
　参考図書 ………………………………………………………………………………… 168
　INDEX …………………………………………………………………………………… 169

化学同人ホームページから音声ファイルと書き込み式問題集をダウンロードできます．
https://www.kagakudojin.co.jp/book/b50174.htm

※音声ファイルの目次は，いちばん最後のページにあります．

本書の特長と効果的な学習法

本書は前半と後半の二つに分かれ，次のようなステップで化学英語の基礎を習得する．

Chapter 1　英語の発音	→	リスニング・スピーキング学習の土台形成
Chapter 2　英単語	→	化学英語論文の読解や作成にそのまま使える型の習得
Chapter 3　頻出基本構文		
Chapter 4　教科書の文章	→	化学英語の文章に慣れ親しむ

音声を最大限活用したリスニング・スピーキングによる学習で基礎をつくる

⬇ Chapter 1～4で学んだ基礎を生かして

Chapter 5　英文和訳	→	化学の英語論文や記事をきっちりとした日本語に訳す
Chapter 6　和文英訳	→	辞書を駆使して化学の英文をつくる
Chapter 7　大学院入学試験	→	学習の総決算

日本語と英語の相互変換

自分で問題を解き、総合的な英語力アップを目指す

本文中のマークなどについて

compound　英単語の下線は代表的なアクセントの位置を示す（アクセントを2か所示した場合は，二重線が第1アクセント，一重線が第2アクセント）．

☐☐　単語や文の前にあるチェック欄は，学習が終わったときなどに使う．

#46　音声が収録されているところにはこのマークが付いている．

Point　学習上注意したい点や問題を解くヒントなどが書いてある．

Jeff's advice　わかりにくい英語用法などを英語監修者がネイティブの観点から指摘している．

　この部分では化学に関する補足解説をしている．

★ 音声ファイルをフル活用しよう

　音声ファイルを最大限活用して効果的に学習しよう．次のような流れで学習すると英語力がメキメキ向上するであろう．

音声ファイルの音声を何度か聴く
↓
テキストを見て意味を確認する
↓
テキストを見ながら音声ファイルを聴く
↓
自分のペースで何度か声に出して読んでみる
↓
オーバーラッピングを数回する
↓
シャドーイングを数回する

● **オーバーラッピングとは**
　テキストを見ながら，音声ファイルの音にかぶせるようにして，一緒に声を出して読む語学学習法．

● **シャドーイングとは**
　音声の英語だけを頼りに，影（shadow）のように後から追いかけて口に出す語学学習法（読まれた 0.5 秒くらい後に追うように読む）．

　オーバーラッピングやシャドーイングでは，慣れないうちは，なかなかスピードについていけないかもしれない．めげずに何度も何度も繰り返し練習しよう．

★ 下敷きを有効活用しよう

　本書に付属している赤い透明の下敷きを使い，単語や文章を労せず覚えてしまおう！　チェック欄もうまく活用しよう！

Keywords

□□ generic	属の	□□ cerium
□□ rare earth	希土類元素	□□ erbium
□□ rich	豊富な	□□ component
□□ ubiquitous		□□ neodymium
	いたるところにある，遍在する	□□ dysprosium
□□ virtually	事実上	□□ laser crystal
□□ costly	（値段が）ひどく高い	（レーザー光発生
□□ do exist		□□ intriguing

★ **辞書について**

英語の学習には適切な辞書が絶対に必要である．

英単語はまず，リーダーズ英和辞典かジーニアス英和大辞典などの大きな辞書で調べよう（ジーニアス中辞典など専門用語がほとんど掲載されていない辞書の使用は推薦できない）．これらの辞書に載っていない単語は，PIDC を使った英辞郎で検索してみよう．

化学に関する英文を書くときは，英和・和英辞典のほか，化学用語辞典，英和活用辞典，英辞郎など，あらゆる情報源を利用して最適な英文を書こう．

◎**英和辞典**

『リーダーズ＋リーダーズプラス英和辞典』(研究社)

リーダーズは 27 万語収録の定評のある万能英和辞書．リーダーズプラスは専門用語も多く掲載されている．

『ジーニアス英和大辞典』(大修館書店)

25 万 5 千語収録．化学用語もかなりカバーしている．文例・用例・解説が豊富で充実している．

『英辞郎』(アルク)

PC にインストールして検索する，英和と和英の両機能を備えた辞書．100 万語収録であらゆる分野の語句が登録されている．PDIC（パーソナルディクショナリー）というブラウザで検索できる．文例が豊富で英文を書くときには重宝する．Web 版は http://www.alc.co.jp/

◎**電子辞典**

英和・和英辞典，英英辞典，広辞苑など複数の辞書が収録されている携帯用電子辞典が販売されている．化学に関する資料を読みこなすには，「リーダース英和辞典」と「ジーニアス英和大辞典」がともに収録されているものが，きわめて便利である．① すばやく適訳が抽出できる，② 複数の辞書を用いて短時間で例文や用法が検索できる，③ ヒストリー機能（英単語帳をつくるのと同じ）がある，④ 携帯に便利である，⑤ 指先を使うだけで検索できる，などの点で，語学学習には今や必携のアイテムである．

◎**化学用語辞典**

『化学英語の活用辞典 第 2 版』(化学同人)

和英と英和の部に分かれており，論文などの英文を練るときには，非常に重宝する 1 冊である．

Chapter 1
発音の基本から始めよう

「日本人の英語の発音はいかにも日本人ぽくて，ネイティブとはほど遠い」と多くの日本人が感じている．それはいったいなぜだろう？ ほとんどの読者は，中学1年の1学期に英語の発音について少し教えられて以来，真剣にこのことについて考えたことがないのではなかろうか．化学英語の学習を始めるにあたって，まず発音を「科学的に」考えてみよう．なぜなら，次のChapter 2で学ぶように，多くの化合物名を発音する際，発音に関する基礎知識が絶対に必要とされるからである．

1.1 音の構成から見た日本語と英語

まず，日本語と英語の「音」の構成の違いを十分に理解し認識することが，英語の上達には不可欠である．たとえば，日本語の「けみすと」と英語の"chemist"を例にとって考えてみよう．日本語の「け・み・す・と」では，四つの音はすべて子音（息がのどの奥，口の閉鎖，舌と歯による閉鎖などにより妨げられながらでる音）と母音（息がほとんど妨げられずにのどからでる音）との組合せから成っている．母音および「ん」を除き，日本語の音は，「子音＋母音」のセットが音の基本構造となる（子音だけを発声することはない）．一方，英語の[kemist]の場合，[ke]と[mi]の部分は「子音＋母音」から成るのに対して，[s]と[t]は子音のみから成っている．また，「け」と[ke]，あるいは「み」と[mi]とを比較すると，英語は子音の部分が強いばかりではなく，日本語とは異なる方法（のど・舌・口の動き――後述）で発せられている．

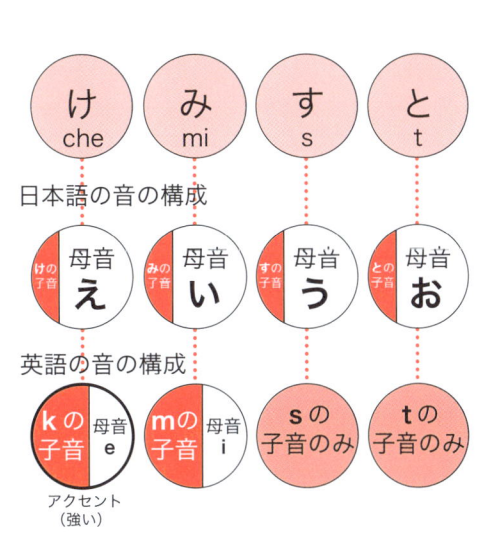

1.2　発音の基本

英語の音を最小単位まで分解すると，数としてはせいぜい 50 個程度しかない．また [t] と [d] など，ほとんど同じ舌と唇の形で無声音（声帯が震えない：voiceless consonant）と有声音（声帯が震える：voiced consonant）の違いで読み分けるもの，あるいは [t] と [l] のように舌の位置がかなり近いものがあり，マスターすべき要素はそれほど多いわけではない．ただし，口の開け方，舌の使い方，のどの使い方が日本語とはかなり異なるので，繰り返し練習する必要がある．管楽器の音をだすのに，最初は誰もが苦労するのと同様，正しい方法で練習しなければ，英語らしい音をだせるようにはならない．自分の口・のど・舌・唇が，各人固有の管楽器であると考え，基本的な音をだす練習を繰り返そう．また，たとえ個々の音を正しくだせたとしても，楽曲がすらすらとすぐには演奏できないのと同様，英文を繰り返し繰り返し読む練習をしないかぎり，音の連続からなる「英語」をすらすら読める（話せる）ようにはならない．綺麗な英語は無理だとしても，相手に通じる英語を話すためには，最低限のルールは守って発音する必要がある．

日本語の母音は「あ・い・う・え・お」の五つであるが，英語の母音は 26 個程度ある（数え方により異なる）．また，日本語の子音は 18 個であるのに対し，英語の子音は，全部で 24 個程度ある．しかし，破裂音（plosive：息の流れをのどなどでせき止めてプレッシャーをかけ，それを解放するときにだす子音）や摩擦音（fricative：息の流れを，唇を噛むなどして口の一部と摩擦させながらだす子音）など，日本語の言葉と比較すると，声というより，機械的な「音」（obstrunt：妨げ音）そのものが，言葉のより重要な構成要素として使われるところに大きな違いがある．ここでは，基本となる母音と子音の音のだし方について，それぞれ学習しよう．

1.3　母音の発音

まず鏡を用意し，光の向きを調節して（鏡を下に向け，反射した室内灯の光で口の中を照らし），自分の口の奥の部分が見えるようにしてみる．そして（普通に）「あーーー」と口を開けたままで言ってみよう．発音すると，舌が下にさがり，口蓋垂（のどちんこ）が見える（あるいは上に上がる）ようになるはずである．次に，口をもう少し大きく開けて，のどに少し力を入れて（あくびをするときのように）口蓋垂をもち上げる（気道を開ける）練習をしよう．続いて，そのままのどの奥を大きく開けた状態で「アアアーーー」，「オオオーーー」と言う練習をしてみよう．英語の母音は，全般に日本語の母音に比べて，このように口蓋垂をもち上げ，喉の奥を大きく開けて，音を響かせながら発音するものが多い．

▶ [ɑ]と[ɑ:]

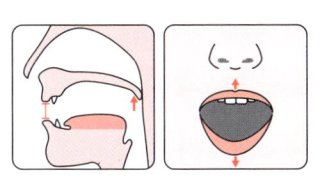

つづりのoにアクセントがあると，[ɑ]の発音になることが多い（ただしイギリス英語では後述する[ɔ]になる）．❶指が縦に2本分入る程度に大きく口を開け，❷口蓋垂をもち上げ「ア」と発音すると[ɑ]になる．[:]は長音記号と呼ばれ，音を伸ばすことを意味する．

#1 [ɑ]

- □□ object　　物体
- □□ compound　　化合物
- □□ hot　　熱い

[ɑ:]

- □□ calm　　平穏な
- □□ palm　　てのひら
- □□ balm　　香油

Point object については，[k][t]の連続した子音の音に注意しよう．

Point compound については，後述する[k][p][d]の三つの破裂音が，クリアーに自分の発音した音のなかに混じっていることを確認しよう．

ただし，単語がeで終わる場合は，oにアクセントがあっても（後述する二重母音）[ou]となる．

- □□ note　　ノート
- □□ scope　　範囲

▶ [æ]

つづりのaにアクセントがあると，[æ]の発音になることが多い．Chapter 2 で述べる alkane, alkene, alkyne のはじめのaが，化学単語のなかでの好例であろう．[æ]は，「エ→ァ」と舌および音をすばやく変化させながらだす，「ア」と「エ」の中間の音である．はじめ「エア」と読んで，だんだん短くする練習をするとよいだろう．あるいは，「ア」から「エ」，逆に「エ」から「ア」に連続的に音を変化させ，中間の音を探すとよい．後述する[i]と同じように口を横に引っ張り，日本語の「え」よりも舌の位置を前にすると発音しやすい．

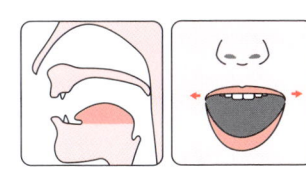

#2 [æ]

- □□ alkane　　アルカン
- □□ gas　　ガス，気体
- □□ halide　　ハロゲン化物
- □□ organic　　有機の

ただし，単語がeで終わる場合は，aにアクセントがあっても[ei]となる．

- □□ grade　　等級，成績
- □□ rate　　割合，速度

▶ [ʌ]

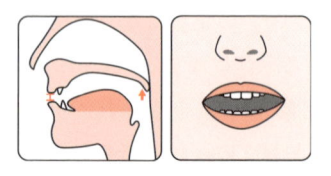

つづりの u にアクセントがあると，[ʌ]の発音になることが多い．突然思いだしたときの「あ！しまった」の「あ」のように，❶ 口は大きく開けずに少しだけ開ける程度で，❷ 口蓋垂をもち上げ，❸ 短く「ア」と発音する．

#3 [ʌ]

| □□ but | しかし | □□ cut | 切る |
| □□ result | 生じる | □□ drug | 薬 |

ただし，単語が e で終わる場合は，u にアクセントがあっても[júː]となる．

| □□ tube | チューブ | □□ produce | 生じる，生産する |

▶ [ər]と[əːr]

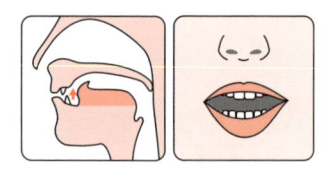

アクセントのない「母音＋r」の組合せのとき，[ər]の音になる．一方，[əːr]は[ər]の長音で，アクセントのある「母音＋r」の組合せで，こちらは強く発音する．後述する子音の[r]のときと同じように，❶ 舌の根本を立てて口を少し開け，❷ 弱く短く「ア」というと[ər]，強く長く「アー」と言うと[əːr]となる．これらは，r-colored vowel と呼ばれる音であるが，アメリカ英語に特徴的で，イギリス英語ではほとんど用いられない．

#4 [ər]　　　　　　　　　　　　　　[əːr]

□□ doctor	博士	□□ early	初期の
□□ air	空気	□□ first	最初の
□□ polar	極性をもつ	□□ furnish	与える
□□ percent	パーセント	□□ term	用語

▶ [ɑːr]

[ɑ]の音から，[ər]の音へ移行させる．したがって，はじめは，❶ 口を縦に大きく開いて[ɑ]から始め，❷ 舌の根本を立てて口を閉じながら[ər]の音へ変化させよう．

#5 [ɑːr]

| □□ heart | 心臓 | □□ start | 開始する |
| □□ large | 大きい | □□ target | 標的 |

1.3 母音の発音

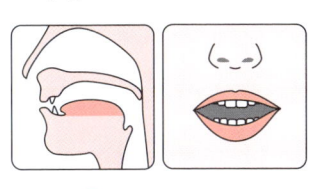

▶ [ə]

アクセントのない母音で「あいまい母音」と呼ばれる．唇，舌に力を入れないで，短く弱くその綴り字本来の母音を弱く発音した音をだし，そこに母音があることを示す．かなり広い範囲の音が含まれる．

#6 [ə]

| □□ about | およそ | □□ permanent | 終身の |
| □□ aluminum | アルミニウム | □□ succeed | 成功する |

▶ [i] と [iː]

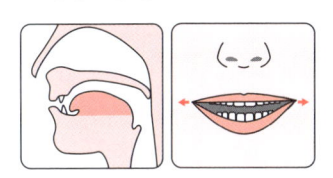

子供がしかめっ面をして，嫌いな相手に「いーだ」というときのように，通常の日本語の「い」より，❶ のどの奥の空気の通り道を狭めて，❷ 高いトーンの音をだすような感じで（あるいは，日本語の「え」よりも，さらに横に口を引っ張り，かつ舌を前にだし，舌の左右の縁の部分を弱く噛むような感じで）「イ」と言うと [i] になる．一方，[iː] は [i] の長音で，「イー」と言う．

#7 [i]　　　　　　　　　　　[iː]

□□ bit	ビット	□□ beat	打つ
□□ fill	満たす	□□ feel	感じる
□□ pick	突く	□□ peak	ピーク
□□ hit	打つ	□□ heat	熱

▶ [u] と [uː]

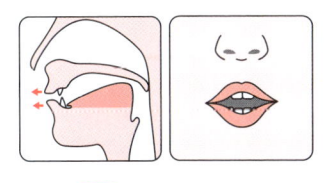

日本語の「う」よりも，❶ のどの奥の空気の抜け道を大きくして，❷ 口を少し突きだし短く「ウ」とすると [u]，口をもっと突きだし「ウー」とすると [uː] になる．

#8 [u]　　　　　　　　　　　[uː]

□□ book	本	□□ improve	改良する
□□ full	いっぱいの	□□ group	集団，基
□□ push	押す	□□ route	経路
□□ look	見る	□□ solution	溶液

▶ [ɛ]（ジーニアス表記は[e]）

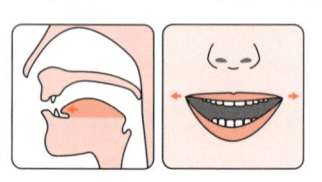

❶日本語の「え」より若干口を横に引っぱることに伴い，❷相対的に舌が前にでた状態で「エ」と発音する．日本語の「え」より少し高めのトーンにする．

#9 [ɛ]

□□ energy　　　　エネルギー　　　　□□ separate　　　分離する
□□ entropy　　　　エントロピー　　　　□□ attempt　　　試みる

▶ [ɔ]と[ɔː]

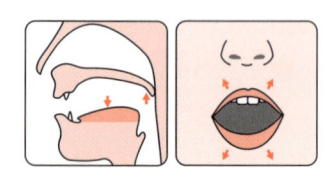

[ɔ]は口を大きく開けて「オ」と喉の奥で響かせる音で，アメリカ英語で[ɑ]のところをイギリス英語では[ɔ]を使う．[ɑ]と同じように，❶のどの奥を空けて（あごと舌を下げ，口蓋垂をもち上げて）なおかつ，❷舌の位置を[ɑ]よりも下げて，口の中の空間をより大きくする．[ɔː]はアメリカ英語でも使われ，[ɔ]の長音で「オー」と高いトーンの音をだす．

#10 [ɔ]　　　　　　　　　　　　　　　[ɔː]

□□ compound（イギリス英語）化合物　　□□ salt　　　　塩
□□ object（イギリス英語）　物体　　　　□□ cause　　　原因

▶ [ɔːr]

[ɑːr]が[ɑ]の音から，[ər]の音へ移行させたのと同様に，[ɔːr]では，[ɔ]から舌を巻き上げて[ər]に移行させる．

#11 [ɔːr]

□□ warm　　　　暖める　　　　□□ form　　　形式
□□ record　　　　記録する　　　　□□ order　　　順番，命令

▶ [ai] [ei] [ɔi] [au] [ou]

これらはいずれも二重母音といわれるもので，はじめの音を強く発音し，後の音は弱く発音する．発音するときの注意点は，個々の音の場合と同じである．

#12 [ai]
　　□□ alkyne　　アルキン
　　□□ science　　科学
[ei]
　　□□ alkane　　アルカン
　　□□ reagent　　試薬
[ɔi]
　　□□ boil　　沸騰する
　　□□ soil　　土

[au]
　　□□ outcome　　結果
　　□□ count　　数える
[ou]
　　□□ soap　　せっけん
　　□□ sodium　　ナトリウム

1.4　子音の発音

▶「か」行に関連した[k]と[g]

　鏡を再び用意しよう．❶ 大きく口を開け，舌の奥(根元)の部分をピタッと上あごの奥にくっつける(気道を舌で塞ぐ)練習をする．「つけて−離す」を何度も繰り返し，舌をもち上げる感覚をつかもう．次に，❷ 舌をもち上げた状態でのどに圧力をかけてみよう．続いて，❸ 舌の力を抜くと，「クゥ」と音がでるはずである．この音が子音の[k]である．声をだすという意識は完全に捨てて，「音」をだす意識が重要である．これは「破裂音」に分類される音で，プレッシャーから解放された空気が吐きだされるときに，のどで奏でる音である．

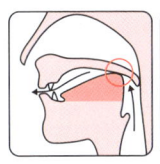

　続いて，口を少しだけ開いた状態で，同じように舌の奥をもち上げ，#13 「クゥ・クゥ・クゥ」と鳴らす練習をしよう．悔しいときに，「クゥッソ〜」とのどの奥で言う際の「ク」の部分だけを取りだした音である．

　続いて，そののどの使い方で #13 「カァ・キィ・クゥ・ケェ・コォ」と詰まらせながら，子音の部分を強調した(妙な)「かきくけこ」をだしてみる．日本語の「かきくけこ」と比較して，母音の発音のときにあごの位置を下げないよう(下げてしまうと，子音の部分がしっかりとでない)，鏡を見ながら何度も何度も繰り返し練習しよう．

　続いて #13 "kick" という語の発音練習をしよう．慣れないうちは，意識して舌をもち上げるようにする．日本語の「き」よりも，のどに圧力がかかっていることを再確認するとともに，子音で音が終わる(「ゥ」の音はださない)ことに注意する．

#13 [k]

□□ keep	続ける	□□ carbon	炭素
□□ ketone	ケトン	□□ cook	料理する

無声音[k]の有声音が[g]である（のどぼとけに手を当てても振動しないのが無声音で，振動するのが有声音）．

[k]と同じようなのどの使い方で #13 「グゥ・グゥ・グゥ」と音をだしてみよう．有声音のときは，「音」をだすというより「声」をだす感覚である．声帯を振動させるコツをつかもう．

続いて，#13 「ガァ・ギィ・グゥ・ゲェ・ゴォ」とのどの奥を狭めて（舌の根本の部分を上あごに当てて）発音する．このときも必要以上にあごの位置を下げないように注意する．

#13 [g]

□□ gold	金	□□ vague	あいまいな
□□ trigger	きっかけとなる	□□ big	大きい

▶ 「さ」行に関連した[s] [θ] [ʃ]

日本語の「さしすせそ」に関連した音は，英語では複数存在する．[s] [θ] [ʃ]は「摩擦音」に分類される．

[s]と[z]

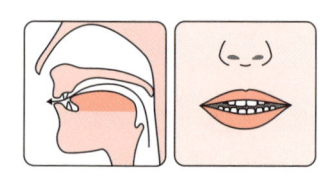

舌と歯茎の間から，息を流すことによる摩擦音．

❶ まず口を軽く開け，上下の歯はくっつくかくっつかないかの距離にし，❷ 舌先を歯の裏に近づけて，#14 「スーーー」と長く摩擦音をだす練習をしよう．少し息を吹いただけで，しっかりと音がでる歯と舌の位置を探してほしい．

続いて，#14 「サァ・シィ・スゥ・セェ・ソォ」としっかりと[s]の摩擦音をだしながら，短い母音を入れた（日本語としては妙な）「さしすせそ」の音をだしてみよう．このときも必要以上にあごを下げないように注意する．

#14 [s]

□□ sea	海	□□ reduce	還元する
□□ silver	銀	□□ lesson	レッスン

[z]は[s]の有声音である．#14 「ズーーー」と濁った摩擦音をだす練習をしよう．のどに手

を当てて声帯が震えていることを確認する．

続いて，#14 「ザァ・ジィ・ズゥ・ゼェ・ゾォ」と音をだしてみよう．このときも必要以上にあごを下げないように注意する．

#14 [z]

□□ zoom　　　　　拡大する　　　│　□□ result　　　　　結果
□□ zinc　　　　　亜鉛　　　　　│　□□ quiz　　　　　　小テスト

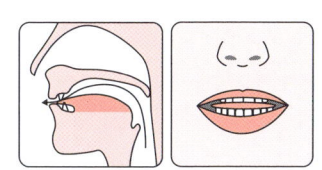

[θ]と[ð]

まず，❶ 上下の前歯が軽く見える程度に口を開く．次に，❷ 舌を歯の位置より前に少しだけだし，❸ 上下の歯で軽く噛む．そのままの状態で，❹ 息を吐き，#15 「スーーー」と摩擦音を長く鳴らす練習をする．thin の場合など，次に母音がくるときは，母音の音に移行させるため舌をすばやく引っ込める動きとなる．息を短く吐きながら，同時に舌をすばやくだしてすばやく引く練習を繰り返そう．

続いて，#15 「サァ・シィ・スゥ・セェ・ソォ」としっかりと [θ] の摩擦音をだしながら音をだしてみよう．やはり，日本語のようにはあごを下げないことに注意する．

#15 [θ]

□□ thin　　　　　薄い　　　　　│　□□ methyl　　　　メチル

this の th は，[θ] の有声音の [ð] である．まったく同じ口の形をつくり，#15 「ズーーー」と声帯を震わせながら，摩擦音をだす練習をしよう．

続いて，#15 「ザァ・ジィ・ズゥ・ゼェ・ゾォ」と母音を入れた摩擦音をしっかりだしてみよう．

#15 [ð]

□□ this　　　　　これ　　　　　│　□□ than　　　　　〜よりも

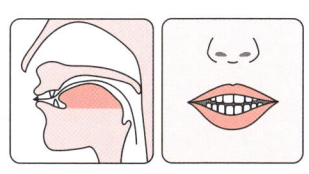

[ʃ]と[ʒ]

人指し指を立て「静かにして」という意味の #16 「シーーー」のように，摩擦音をだす．[s] よりも唇を突きだす．このとき，口先を少しすぼめて，上下の唇の間に間隔を開けるようにすると，この音をだしやすい．

続いて，#16 「シァ・シィ・シゥ・シェ・シォ」と音をだしてみよう．やはり，日本語のようにあごを下げすぎないよう注意する．

Chapter 1 発音の基本から始めよう

#16 [ʃ]
- □□ she 彼女
- □□ sugar 糖
- □□ reaction 反応
- □□ flash 閃光

声帯を鳴らし，#16 「ジーーー」とすると有声音の[ʒ]になる．[ʃ]の有声音である．

続いて，#16 「ジァ・ジィ・ジゥ・ジェ・ジォ」と音をだしてみよう．やはり，日本語のようにあごを大きく下げないように注意する．

#16 [ʒ]
- □□ vision 視力
- □□ conclusion 結果
- □□ decision 決定
- □□ measure 測る

[ʒ]は摩擦音（長く音がだせる）で，後述する[dʒ]は破擦音（一瞬しか音がでない）である．英語では[dʒ]の音が多く，[ʒ]の音は，「〜sion」「〜sure」などかなり限られた語でしか使われない．

▶「た」行に関連した[t] [tʃ]

[t]と[d]

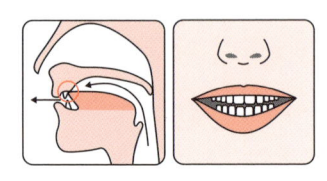

日本語では「子音＋母音」の音が多いのに対し，英語では「子音」で終わる音が多い．代表的な音が[t]である．

まず，❶ 口を軽く開け，舌先を上の歯茎にぴったりとつけてみよう．続いて，❷ 舌先の位置をキープしたままで，息を吐くときの圧力を舌先にかける．次に❸ 舌にかけている力を抜けば，#17 「ッゥー」という音がでるはずである（あるいは逆に，吐いている息を，舌先を歯茎につけることで止める練習をすれば，圧力がかかる感覚がつかめるはずである）．この場合も声をだすという意識を100%捨てて，物理的な力だけで音をだすように短く強く息を吐いてみよう．これは，舌先と歯茎の間にかけた圧力を解放するときに生じる「破裂音」である．また，この音は声帯を震わせずに音をだす無声音でもある．

続いて，日本語の「た・ち・つ・て・と」の母音の部分をカットする意識で，短く #17 「タァ・チィ・ツゥ・テェ・トォ」という練習を繰り返そう．同じく，あごを下げすぎないよう注意する．

#17 [t]
- □□ tablet 錠剤
- □□ table 表
- □□ attend 出席する
- □□ best 最高の

[d]は[t]の有声音である．

[t]とまったく同じように舌先を上の歯裏にセットして，声帯を震わせて，#17 「ダァ」と鳴らしてみよう．

続いて，短く #17 「ダァ・ディ・ドゥ・デェ・ドォ」と発音練習してみよう．この場合も母音部分をカットする意識をもつとよいだろう．やはり，あごを下げすぎないように注意する．また，[t]や[d]で終わる単語のときに母音が入らないようにも注意する．

#17 [d]

☐☐ dip　　　　ちょっと浸ける　　☐☐ needle　　針
☐☐ dry　　　　乾かす　　　　　　☐☐ good　　　良い

[tʃ]と[dʒ]

つまらないときやがっかりして「ちぇ」と言うときの「ち」の音を，口をもっと閉じ気味に，なおかつ「ぇ」の音をかなり取り除いてだしてみよう．

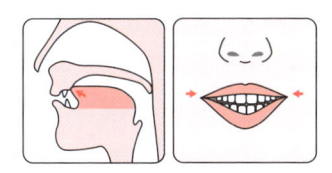

❶ [ʃ]の口と舌の状態からスタートし，❷ 舌を上歯茎に軽く接触させ（[t]のときよりも少し奥），❸ 息を吐いて短く #18 「チェ」「チェ」「チェ」と数回繰り返してみる．破裂音[t]と摩擦音[ʃ]が組み合わさった一つの音で，破擦音(affricate)と呼ばれる．

次に #18 「チァ・チィ・チゥ・チェ・チョ」と短く母音を入れて言ってみよう．

#18 [tʃ]

☐☐ check　　　調べる　　　　　☐☐ bleach　　　漂白する
☐☐ charge　　 充電　　　　　　☐☐ mixture　　 混合物

[d]が[t]の，また，[ʒ]が[ʃ]の有声音であったのと同様に，[dʒ]は[tʃ]の有声音である．

[tʃ]の口と舌をまずセットする．その状態で声帯を震わせ #18 「ジェ」「ジェ」「ジェ」と鳴らそう．さらに #18 「ジァ・ジィ・ジゥ・ジェ・ジョ」と発音してみよう．

#18 [dʒ]

☐☐ jack　　　　ジャッキ　　　　☐☐ subject　　 学科
☐☐ jargon　　　専門用語　　　　☐☐ stage　　　 段階

[ʒ]とは異なり[dʒ]は，舌を上歯茎につけて始める破擦音であり，一瞬で終わる．しかし，実際に二つの音を区別するのは難しい．

▶「な」行および「ん」に関連した[n]と「ま」行に関連した[m]，および[ŋ]

[m], [n], [ŋ]はいずれも鼻に息を通しながらだす有声音の子音で，鼻音（nasal）に分類される．

[m]

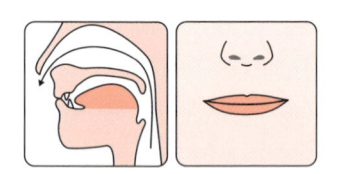

まず口を閉じたまま，息を吐いてみよう．いわゆる鼻息であるが，強く吐いても大きな音量はでず，言葉として英語には用いられない．一方，同じように，❶鼻から息を抜きながら，❷声帯を鳴らす（唸る）と子音の[m]となる．すなわち[m]は，これまで学習した子音とは異なり，有声音しかない子音である．

次に，#19 「m＿＿＿」と唸りながら「マァ」と言ってみる．さらに，#19 「m＿＿マァ・m＿＿ミィ・m＿＿ムゥ・m＿＿メェ・m＿＿モォ」と数回繰り返してみよう．

唸る感覚がつかめてきたら，はじめに短く唸りの音を混じらせて，#19 「mマァ・mミィ・mムゥ・mメェ・mモォ」と練習しよう．[m]の音は子音なので，語尾に来るときは日本語の「む」と異なり，口を最後に開いて母音を入れてはいけない．

#19 [m]
- □□ molecule　分子
- □□ mechanism　メカニズム
- □□ some　いくつかの
- □□ aluminum　アルミニウム

[n]

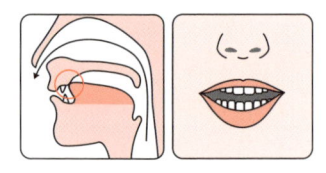

一方[n]は，❶口を少し開け，❷舌先を上の歯茎（[t]よりも若干奥）にぴったりとつけて，❸ #19 「ンーー」と唸ると[n]の音になる（日本語の「な・に・ぬ・ね・の」の出だしでも，無意識のうちに舌先を上の歯茎奥に軽くつけていることを確認しよう）．続いて，#19 「n＿＿ナァ・n＿＿ニィ・n＿＿ヌゥ・n＿＿ネェ・n＿＿ノォ」と言ってみよう．

さらに，短く唸りの音を混じらせて，#19 「nナァ・nニィ・nヌゥ・nネェ・nノォ」と練習しよう．

#19 [n]
- □□ number　数
- □□ once　一度
- □□ contain　含む
- □□ clean　清潔な

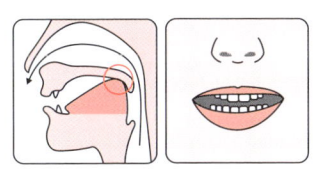

1.4 子音の発音

[ŋ]

また[ŋ]については，❶ [k]あるいは[g]と同じように舌の奥（根本）の部分をピタッと上あごの奥につける．❷ その状態で鼻から息を抜き #19 「ンー」と唸ってみよう．それが[ŋ]の音になる．[n]+[g]ではないことに注意．

#19 [ŋ]

□□ lo<u>ng</u>	長い	□□ engineeri<u>ng</u>	工学
□□ stro<u>ng</u>	強い	□□ si<u>ng</u>let	シングレット

▶「は」行に関連した[f] [h] [p]

[f]と[v]

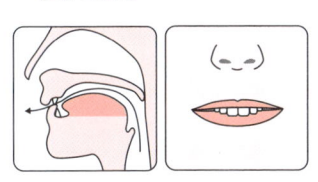

まず，❶（はじめてこの音の練習をする場合は）下唇を上下の前歯でしっかりと挟んで，❷ そのままの状態で声をだす意識をもたずに息を吐き，❸ 上の歯と唇の間で起こる摩擦音（長い時間だせる）をだす感覚を会得しよう． #20 「フゥーーー」と長く音を発生させてみよう．日本人にとっては雑音としか思えないような摩擦音をしっかりだすこと．

次に，アルファベットの #20 F を発音してみよう．日本語の「ふ」のように母音の「う」が入らないように注意．

これに慣れてきたら，下唇の内側の部分に軽く上の前歯を当て，同じように #20 「フゥーーー」と長く，摩擦音をだす練習をしよう．

続けて短く #20 「ファ・フィ・フゥ・フェ・フォ」と，子音の部分を強調した「は・ひ・ふ・へ・ほ」の発音練習をしてみよう．やはりあごを下げすぎないようにすることが母音の部分をカットするコツである．

#20 [f]

□□ <u>f</u>act	事実	□□ <u>f</u>ifth	第5の
□□ <u>f</u>ast	速い	□□ af<u>f</u>ord	与える

[v]は[f]の有声音である．

同じように，歯と唇をセットして， #20 「ヴゥーーー」と声帯を長く震わせる練習をしよう．続いて，短く #20 「ヴァ・ヴィ・ヴゥ・ヴェ・ヴォ」と，子音の部分を強調した（母音をカットした）

「ば・び・ぶ・べ・ぼ」の発音練習をしてみよう．

#20 [v]

☐☐ vinegar	酢	☐☐ save	保存する
☐☐ vast	広大な	☐☐ device	装置

[h]

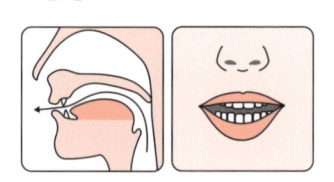

寒くて手がかじかんだときに #21 「はぁー」と息を吹きかけるように，のどの奥に息を通す摩擦音である．

まず，声帯をほとんど震わせずにのどの奥で #21 「ハァ・ヒィ・フゥ・ヘェ・ホォ」と鳴らしてみよう．この音が語尾に来ることはなく，常に「子音＋母音」で使われる音（[h]自体は無声音）である．「母音」の前にはっきりと摩擦音が混じるように発音してみよう．

#21 [h]

☐☐ hold	保つ	☐☐ hint	ヒント
☐☐ hacker	ハッカー	☐☐ heat	熱

[p]と[b]

上下の唇を使った破裂音（一瞬しか音がでない）である．

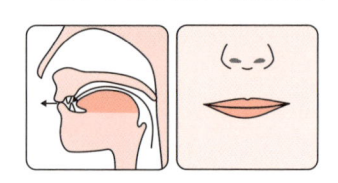

まず，❶ 唇を閉じた状態で，のどから唇に圧力をかける．続いて，❷ 唇の力を抜いてみよう． #22 「プゥ」と音がでるはずである．このとき，唇が息による圧力で（あごを動かすことなく）「プゥ」と自然に開けられる感覚をつかむまで練習する．

次に，日本語の「ぱ・ぴ・ぷ・ぺ・ぽ」の母音の部分をカットして，短く #22 「パァ・ピィ・プゥ・ペェ・ポォ」と無声音をはっきりとだした「ぱ・ぴ・ぷ・ぺ・ぽ」の練習を繰り返そう．やはりこのとき，あごを大きく下げてしまうと，綺麗な破裂音がでない．[t]の場合と同様，無声音である[p]で終わる単語のときには，「ぷう」と母音を入れないように注意する．

#22 [p]

☐☐ pen	ペン	☐☐ expect	期待する
☐☐ possible	可能な	☐☐ step	段階

[b]は[p]の有声音である．「ば・び・ぶ・べ・ぼ」の母音の部分をカットして，短く #22 「バァ・ビィ・ブゥ・ベェ・ボォ」と破裂音であることが感覚としてつかめるまで繰り返し練習しよう．

1.4 子音の発音　15

#22 [b]

□□ bond	結合	□□ member	一員
□□ butane	ブタン	□□ tab	つまみ，ラベル

▶「や」行に関連した[j]と「わ」行に関連した[w]

[j][w]はともに，舌あるいは口の形を変化させながら音をだす「移行音」であり，単独で母音に近い音なので半母音（semivowel）と呼ばれる（ただし，これらの音で始まる単語の前に不定冠詞を置くときは，an ではなく a を用いることに注意）．

[j]

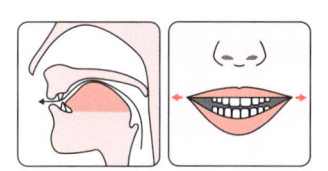

[j]の発音ではまず，❶唇を左右に少し広げ，❷舌の前の部分を広く上あごにつけ，発音の準備をする．次に，❸舌を上あごから離し，[i]の位置にすばやく移行させながら **#23**「イ」「イ」「イ」「イ」と何度か言ってみよう．舌とのどに少し力を入れて発音しよう．そのとき鏡で舌の動きを確認すること．

次に，**#23**「イァ・イィ・イゥ・イェ・イォ」と母音に移行する練習をしよう．これに対して[i]は，舌を上にもち上げずそのまま「イ」の音をだす．

#23 [j]

□□ yield	収率	□□ yeast	酵母
□□ yet	まだ	□□ opinion	意見

[w]

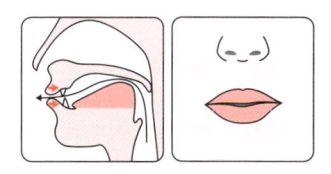

[w]は，❶唇をできるだけ突きだし，❷力を抜いて突きだした唇を元にもどしながら，❸ **#23**「ウ」と音をだす．

続いて，**#23**「ウァ・ウィ・ウゥ・ウェ・ウォ」と，突きだした口を次の母音の形に移行させながら発音練習をしてみよう．

次に **#23** "woman"という語の読みを練習しよう．[w]と[u]の2種類の「ウ」が連続していることに注意してほしい（「ウーマン」ではない）．

#23 [w]

□□ would	will の過去	□□ quick	すばやい
□□ waste	廃棄物	□□ twenty	20

▶「ら」行に関連した [l] と [r]

英語の [l] と [r]，日本語の「ら」行の子音の発音の大きな違いは舌の位置にある．

[l]

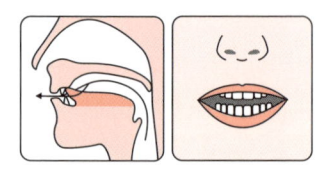

まず鏡をもって口を大きく開き，光の加減を調節して上の歯全体を映してみる．次に，❶ 舌を上にもち上げて，舌先を上の歯茎にぴったりとつける．[l] の発音では，舌をこの位置にセットすることが基本となる．そのうえで，❷ 唇を横に若干引っ張りながら，上の歯の半分と，下の歯の 70% ぐらいが鏡に映る程度に口を開く．歯と歯の間は若干隙間をつくる．その状態で，❸ #24 「ラ」と舌の両側から発音する．[l] から始まる語は，この舌の位置から読み始め，先頭以外にある場合は，この状態に移行するように舌と口の動きを入れて「ラ」と発音する．

#24 [l]

□□ a<u>l</u>lyl(C_3H_5-)　アリル　　　　□□ g<u>l</u>ass　　ガラス
□□ <u>l</u>ight　　光　　　　　　　　　□□ p<u>l</u>ay　　遊ぶ

[r]

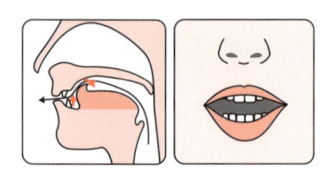

一方，[r] の発音の仕方にはいくつかあるようだが，一つの方法は，❶ 若干口先をすぼめ，❷ 舌の根本をもっと立て，のどの奥のほうに舌を丸め込んで，❸ #24 「ラ」と言う．音がこもる最適な舌の位置を探そう．

#24 [r]

□□ a<u>r</u>yl(Ar-)　アリール　　　　　□□ g<u>r</u>ass　　草
□□ <u>r</u>ight　　正しい　　　　　　　□□ p<u>r</u>ey　　えじき

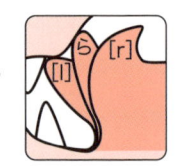

なお，日本語の「ら」行の子音は，英語の l と r の間で口の中心ぐらいの上あごに舌を（われわれは無意識のうちに）軽くつけて発音している．英語の [l]，日本語の「ら」，英語の [r] を順番に読み分け，それぞれの違いを舌および口に覚え込ませよう．[r] を練習するときは，小さく「ゥ」を加え，「ゥラ」と発音するのもよい．どの程度舌を立てるかは，次にくる母音の種類によっても異なり，また個人差もかなり大きいようである．

1.5 口の形から見た発音のまとめ

　ここまでは，日本語の50音と比較して英語の発音について考えてきたが，ここでは，舌，唇，あごなどハードの側面から，日本語と英語の音の違いを分類してみよう．

▶ 舌の位置から音を分類

　まず舌の位置から日本語と英語の違いについて考えてみたい．

　日本語の50音を順番に発音してみると，「あ」行と「か」行では，舌の位置をほとんど動かさずに発音している．「さ」行では歯の裏に舌を近づけて発音している．「た」行では，舌先を歯茎から若干奥の部分を(無意識のうちに)軽く接触させて発音していることを確認してほしい．また「な」行では，「た」行より若干奥の部分に，「ら」行では「な」行よりさらに奥の部分に軽く舌先を接触させて，日本語の子音の部分を発音している．「は」行と「や」行では，「あ」行などと同じく，あまり舌の位置を動かさずに発音している．

　一方，英語では，日本語よりも舌の位置の移動をより大きく行う．最も舌をだすのが[θ]で，舌を歯の位置より前にもってくる．[s]では，日本語の「さ」行とほとんど舌の位置は変化がないが，日本語と比べ，上の歯と下の歯の間隔をもっと狭めて，息を吐くことで摩擦音であることをより明確に示す．日本語の「た」行よりもっとぴったりと舌先を歯茎につけて発音するのが，[t]と[l]である．それよりも軽く舌先をつけるのが[tʃ]．日本語の「な」行や「ら」行と同じくらい，口の天井に舌先をつけて発音するのが[n]と[j]である．一方，舌をもっと巻き上げて，のどの奥の部分に舌先を近づける音には[əːr] [ɑːr] [ɔːr] [r]がある．また，[k]は日本語の「か」行よりもさらに舌の根本の部分をもち上げ，のどの奥の空気の通り道をふさいでだす．

▶ 唇の位置から音を分類

　[ɑ] [ɔ] [i] [ɛ] [æ] [u] [h] [s] [ʃ] [f] [p] [w]などでは，舌の位置よりも，唇の位置(あごや歯の位置)が日本語と大きく異なる．はじめにも述べたように，日本語は「小さな子音＋大きな母音」の構造が基本であるため，それぞれの音の終わりは，「ん」を除き「あ・い・う・え・お」いずれかの唇の形で終わる．しかし，英語ではもっと唇の動きのバリエーションが大きい．[w]は大きく唇を突きだす．[u] [h] [ʃ] [tʃ]では若干唇を突きだし，唇の間隔を開けるとそれらしい音をだしやすい．[ɑ] [ɔ]では，上下の唇の間隔を大きく開ける(つまり，あごを大きく下げる)ことに注意したい．また，[i] [ɛ] [æ] [l]では，唇を横に若干引っ張るとよい．[f]では，上の歯を軽く，下唇に接触させる．[p]では，あごが下がるのを抑えて，空気の圧力で唇が「ぷうっ」と開けられる感覚をつかむことなどが重要であることも前述したとおりである．

Chapter ❷ 化学の基本単語を覚えよう

> このChapterでは，化学関連の基本単語を覚えよう．Chapter 1で学習した発音に注意して，不明な点があれば辞書で発音記号をチェックしよう．ここでも，オーバラッピングやシャドーイング（巻頭「本書の特長と効果的な学習法」参照）で音声ファイルをおおいに活用しよう．

2.1 化学で重要な接頭語と接尾語

まず，化学でよく用いられる接頭語と接尾語を学習しよう．

2.1.1 接頭語 (prefix)

▶成分比を示すためにはギリシャ語を使う

#25

□□ mono （一つの）	□□ penta （五つの）	□□ nona （九つの）
□□ di （二つの）	□□ hexa （六つの）	□□ deca （10個の）
□□ tri （三つの）	□□ hepta （七つの）	□□ undeca （11個の）
□□ tetra （四つの）	□□ octa （八つの）	□□ dodeca （12個の）

□□ carbon monoxide 一酸化炭素	□□ dipole moment 双極子モーメント
□□ monolayer 単層	□□ triangle 三角形
□□ dimer 二量体	□□ triplet 三重項

Chapter 2 化学の基本単語を覚えよう

▶原子団などの数を示すためには次の数詞を使う

#26

- ☐☐ bis （二つの）
- ☐☐ tris （三つの）
- ☐☐ tetrakis （四つの）
- ☐☐ pentakis （五つの）
- ☐☐ hexakis （六つの）
- ☐☐ heptakis （七つの）
- ☐☐ octakis （八つの）

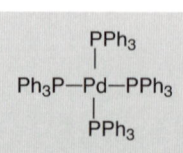

$$Ph_3P\text{—}Pd\text{—}PPh_3$$ （with PPh₃ above and below Pd）

Pd(PPh₃)₄; tetrakis(triphenylphosphine)palladium
トリフェニルホスフィンには，すでに数詞トリ(tri)がついているので，カッコ内に入れ，tetrakis をつける．

#27

▶ bi（バイ）ter（ター）：環および環集合体の繰返し

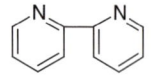

- ☐☐ bipyridine ビピリジン
- ☐☐ terphenyl ターフェニル

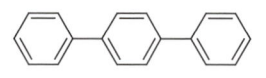

▶ un-：否

- ☐☐ unsaturation 不飽和
- ☐☐ unambiguous 明白な

▶ amph-：両

- ☐☐ amphiphilic 両親媒性の
- ☐☐ amphoteric 両性の

▶ dis-：反

- ☐☐ disagree 反対する
- ☐☐ disadvantage 不利，欠点

▶ in-：内へ

- ☐☐ injection 注射
- ☐☐ inclusion 包接

▶ inter-：間

- ☐☐ intermolecular 分子間の
- ☐☐ intermediate 中間体

▶ intra-：内部の
　□□ intramolecular　分子内の　　□□ intracellular　細胞内の

▶ pseud-：擬似の
　□□ pseudohalogen　擬ハロゲン　　□□ pseudorotation　擬似回転

▶ lipo-：脂肪の
　□□ lipophilic　親油性の　　□□ liposome　リポソーム
　　　　　　　　　　　　　　　　　　　　　　（人工リン脂質小胞）

▶ oligo-：少数の・いくつかの
　□□ oligomer　オリゴマー　　□□ oligonucleotide　オリゴヌクレオチド

▶ re-：再び
　□□ recycling　リサイクリング　　□□ reuse　再利用する

2.1.2　接尾語(suffix)

#28

▶ 〜able：〜できる
　□□ reliable　信頼できる　　□□ biodegradable　生分解性の

▶ 〜fy：〜にする
　□□ liquefy　液化する　　□□ solidify　固化する

▶ 〜phobic：〜嫌いの
　□□ hydrophobic　疎水性の　　□□ acrophobic　高所恐怖症の

▶ 〜philic：〜好きの
　□□ nucleophilic　求核性の　　□□ hydrophilic　親水性の

▶ 〜lysis：〜分解
　□□ electrolysis　電気分解　　□□ hydrolysis　加水分解

2.1.3 乗　数

#29

☐☐ 10^{-15}	femto	☐☐ 10^{-3}	milli	☐☐ 10^{2}	hecto
☐☐ 10^{-12}	pico	☐☐ 10^{-2}	centi	☐☐ 10^{6}	mega
☐☐ 10^{-9}	nano	☐☐ 10^{-1}	deci	☐☐ 10^{9}	giga
☐☐ 10^{-6}	micro	☐☐ 10	deca	☐☐ 10^{12}	tera

2.2　元素の名称

元素記号の語源には，ドイツ語あるいはラテン語由来のものがあるため，元素の英語名との相関がわかりにくいものがいくつかある．たとえば，鉄の元素記号は Fe であるが，これはラテン語の ferrum に由来し，英語の iron とはまったく異なる．ここでもシャドーイング，オーバーラッピングするなど，音声ファイルを上手に活用して効率よく学習しよう．

#30

☐☐ H	hydrogen	水素	☐☐ Al	aluminum	アルミニウム
☐☐ ^{1}H	protium	プロチウム	☐☐ Si	silicon	ケイ素
☐☐ D	deuterium	重水素	☐☐ P	phosphorus	リン
☐☐ T	tritium	トリチウム	☐☐ S	sulfur（英 sulphur）	硫黄
☐☐ ^{1}H^{+}	proton	陽子	☐☐ Cl	chlorine	塩素
☐☐ D^{+}	deuteron	重陽子	☐☐ Ar	argon	アルゴン
☐☐ T^{+}	triton	トリトン	☐☐ K	potassium	カリウム
☐☐ He	helium	ヘリウム	☐☐ Ca	calcium	カルシウム
☐☐ Li	lithium	リチウム	☐☐ Sc	scandium	スカンジウム
☐☐ Be	beryllium	ベリリウム	☐☐ Ti	titanium	チタン
☐☐ B	boron	ホウ素	☐☐ V	vanadium	バナジウム
☐☐ C	carbon	炭素	☐☐ Cr	chromium	クロム
☐☐ N	nitrogen	窒素	☐☐ Mn	manganese	マンガン
☐☐ O	oxygen	酸素	☐☐ Fe	iron	鉄
☐☐ F	fluorine	フッ素	☐☐ Co	cobalt	コバルト
☐☐ Ne	neon	ネオン	☐☐ Ni	nickel	ニッケル
☐☐ Na	sodium	ナトリウム	☐☐ Cu	copper	銅
☐☐ Mg	magnesium	マグネシウム	☐☐ Zn	zinc	亜鉛

2.2 元素の名称

☐☐	Ga	gallium	ガリウム	☐☐	Xe	xenon	キセノン
☐☐	Ge	germanium	ゲルマニウム	☐☐	Cs	cesium	セシウム
☐☐	As	arsenic	ヒ素	☐☐	Ba	barium	バリウム
☐☐	Se	selenium	セレン	☐☐	La	lanthanum	ランタン
☐☐	Br	bromine	臭素	☐☐	Hf	hafnium	ハフニウム
☐☐	Kr	krypton	クリプトン	☐☐	Ta	tantalum	タンタル
☐☐	Rb	rubidium	ルビジウム	☐☐	W	tungsten	タングステン
☐☐	Sr	strontium	ストロンチウム	☐☐	Re	rhenium	レニウム
☐☐	Y	yttrium	イットリウム	☐☐	Os	osmium	オスミウム
☐☐	Zr	zirconium	ジルコニウム	☐☐	Ir	iridium	イリジウム
☐☐	Nb	niobium	ニオブ	☐☐	Pt	platinum	白金
☐☐	Mo	molybdenum	モリブデン	☐☐	Au	gold	金
☐☐	Tc	technetium	テクネチウム	☐☐	Hg	mercury	水銀
☐☐	Ru	ruthenium	ルテニウム	☐☐	Tl	thallium	タリウム
☐☐	Rh	rhodium	ロジウム	☐☐	Pb	lead	鉛
☐☐	Pd	palladium	パラジウム	☐☐	Bi	bismuth	ビスマス
☐☐	Ag	silver	銀	☐☐	Po	polonium	ポロニウム
☐☐	Cd	cadmium	カドミウム	☐☐	At	astatine	アスタチン
☐☐	In	indium	インジウム	☐☐	Rn	radon	ラドン
☐☐	Sn	tin	スズ	☐☐	Fr	francium	フランシウム
☐☐	Sb	antimony	アンチモン	☐☐	Ra	radium	ラジウム
☐☐	Te	tellurium	テルル	☐☐	Ac	actinium	アクチニウム
☐☐	I	iodine	ヨウ素	☐☐	U	uranium	ウラン

	1	2	3	4	5	6	7	8	9	10	11	12	13	14	15	16	17	18
1	H																	He
2	Li	Be											B	C	N	O	F	Ne
3	Na	Mg											Al	Si	P	S	Cl	Ar
4	K	Ca	Sc	Ti	V	Cr	Mn	Fe	Co	Ni	Cu	Zn	Ga	Ge	As	Se	Br	Kr
5	Rb	Sr	Y	Zr	Nb	Mo	Tc	Ru	Rh	Pd	Ag	Cd	In	Sn	Sb	Te	I	Xe
6	Cs	Ba	L	Hf	Ta	W	Re	Os	Ir	Pt	Au	Hg	Tl	Pb	Bi	Po	At	Rn
7	Fr	Ra	A	Rf	Db	Sg	Bh	Hs	Mt									
			L	La	Ce	Pr	Nd	Pm	Sm	Eu	Gd	Tb	Dy	Ho	Er	Tm	Yb	Lu
			A	Ac	Th	Pa	U	Np	Pu	Am	Cm	Bk	Cf	Es	Fm	Md	No	Lr

2.3 化合物の名称

基本的な化合物の読み方をマスターしよう．まず，アルカン，アルケン，アルキンの読みに注意．これらの日本語のカタカナ化合物名は，綴り字から機械的に字訳したものである．実際には，アルカンが「エァルケエィン」，アルケンが「エァルキーン」，アルキンが「エァルカアィン」のように，あたかも日本語と英語の読みが一つずつずれているかのごとく発音される．フォニックス (phonics：綴り字と発音の関係の研究) の点から，Chapter 1 の母音のところでもふれたように，a にアクセントがあると[æ]となること，アルカンの語尾〜 ane の「a」を二重母音[ei]と読み，アルケンの語尾〜 ene の「e」を長母音[iː]と読み，アルカンの語尾〜 yne の「y」を二重母音[ɑi]と読むことを頭に入れれば対応できるであろう．

なお，多くの化合物は慣用名で呼ばれている．

#31

C_nH_{2n+2}
□□ alkane
アルカン

Point
アルカン (〜 ane) の「a」は二重母音[ei]である．

CH_4
□□ methane
メタン

C_2H_6
□□ ethane
エタン

C_3H_8
□□ propane
プロパン

C_4H_{10}
□□ butane
ブタン

C_5H_{12}
□□ pentane
ペンタン

C_6H_{14}
□□ hexane
ヘキサン

C_7H_{16}
□□ heptane
ヘプタン

C_8H_{18}
□□ octane
オクタン

C_9H_{20}
□□ nonane
ノナン

$C_{10}H_{22}$
□□ decane
デカン

#32

C_nH_{2n}

□□ alk__e__ne
アルケン

Point
アルケン（〜ene）の「e」は長母音［iː］である．

C_2H_4
□□ eth__e__ne　　□□ eth__y__lene
エテン あるいは エチレン

C_3H_6
□□ pr__o__pene　　□□ pr__o__pylene
プロペン あるいは プロピレン

C_4H_8
□□ b__u__tene　　□□ b__u__tylene
ブテン あるいは ブチレン

C_5H_{10}
□□ p__e__ntene
ペンテン

C_6H_{12}
□□ h__e__xene
ヘキセン

C_7H_{14}
□□ h__e__ptene
ヘプテン

C_8H_{16}
□□ __o__ctene
オクテン

C_9H_{18}
□□ n__o__nene
ノネン

#33

C_nH_{2n-2}

□□ alk__y__ne
アルキン

Point アルキン（〜yne）の「y」は二重母音［ai］である．

C_2H_2
□□ eth__y__ne　　□□ ac__e__tylene
エチン あるいは アセチレン
（「〜リ〜ン」と発音）

C_3H_4
□□ pr__o__pyne
プロピン

C_4H_6
□□ b__u__tyne
ブチン

C_5H_8
□□ p__e__ntyne
ペンチン

C_6H_{10}
□□ h__e__xyne
ヘキシン

C_7H_{12}
□□ h__e__ptyne
ヘプチン

C_8H_{14}
□□ __o__ctyne
オクチン

C_9H_{16}
□□ n__o__nyne
ノニン

#34

R–OH
□□ alcohol
アルコール（〜ol）

Point アルコールは，対応するアルカンの語尾「〜e」を「〜ol」に置換して命名するのが基本．

MeOH
□□ methanol
メタノール

EtOH
□□ ethanol
エタノール

BuOH
□□ butanol
ブタノール

□□ phenol
フェノール

□□ benzyl alcohol
ベンジルアルコール

□□ ethylene glycol
エチレングリコール

□□ glycerol
グリセロール

□□ glycerin
グリセリン

#35

□□ aldehyde
アルデヒド（〜al）

Point アルデヒドは，対応する同じ炭素数のアルカンの語尾「〜e」を「〜al」に置換して命名するのが基本．

□□ formaldehyde
ホルムアルデヒド
あるいは
□□ methanal
メタナール

□□ acetaldehyde
アセトアルデヒド
あるいは
□□ ethanal
エタナール

□□ butylaldehyde
ブチルアルデヒド
あるいは
□□ butanal
ブタナール

□□ benzaldehyde
ベンズアルデヒド

#36

□□ ketone
ケトン（〜one）

□□ 3-hexanone
3-ヘキサノン

□□ acetone
アセトン

□□ benzophenone
ベンゾフェノン

Point ケトンは，対応するアルカンの語尾「〜e」を「〜one」に置換して命名するのが基本．

2.3 化合物の名称

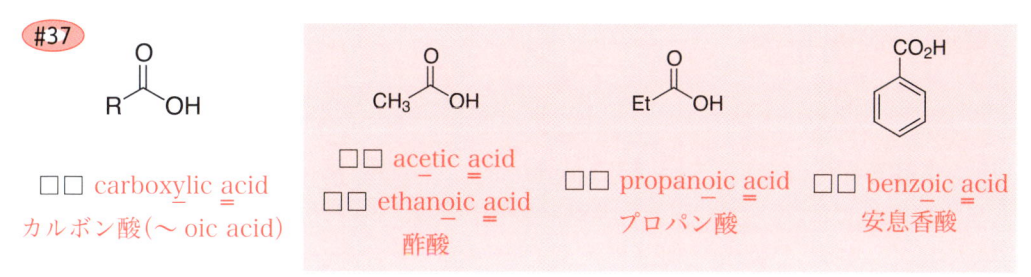

#37

□□ carboxylic acid
カルボン酸（〜 oic acid）

□□ acetic acid
□□ ethanoic acid
酢酸

□□ propanoic acid
プロパン酸

□□ benzoic acid
安息香酸

Point カルボン酸は，対応する同じ炭素数のアルカンの語尾「〜 e」を「〜 oic acid」に置換して命名するのが基本．

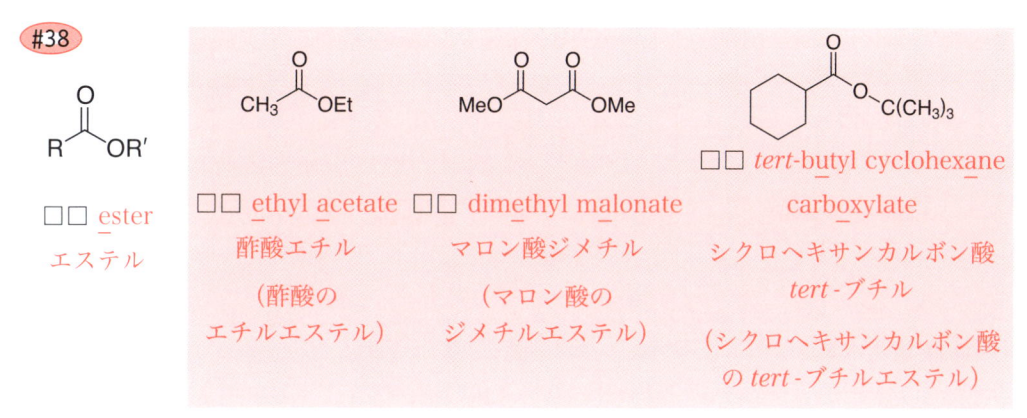

#38

□□ ester
エステル

□□ ethyl acetate
酢酸エチル
（酢酸の
エチルエステル）

□□ dimethyl malonate
マロン酸ジメチル
（マロン酸の
ジメチルエステル）

□□ tert-butyl cyclohexane carboxylate
シクロヘキサンカルボン酸
tert-ブチル
（シクロヘキサンカルボン酸
の tert-ブチルエステル）

Point エステルは，R'のアルキル基名＋対応する酸の語尾「〜 ic acid」を「〜 ate」に変えたもの．

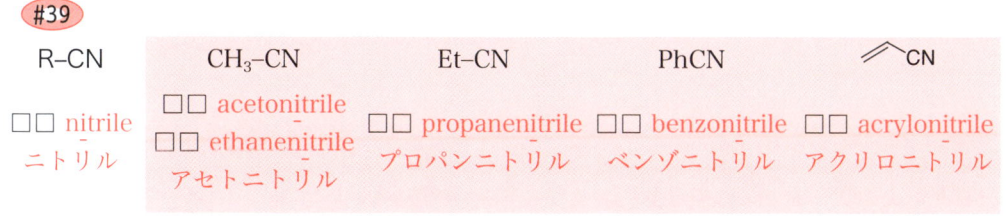

#39

R–CN   CH₃–CN   Et–CN   PhCN

□□ nitrile
ニトリル

□□ acetonitrile
□□ ethanenitrile
アセトニトリル

□□ propanenitrile
プロパンニトリル

□□ benzonitrile
ベンゾニトリル

□□ acrylonitrile
アクリロニトリル

Point ニトリルは，対応する同じ炭素数のアルカンの語尾に「nitrile」をつけて命名するのが基本．

□□ ether
エーテル

□□ acetal
アセタール

□□ amine
アミン

□□ urea
尿素

#40 ヘテロ環化合物

□□ epoxide
エポキシド

□□ pyridine
ピリジン

□□ pyrrole
ピロール

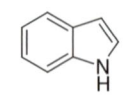

□□ thiophene
チオフェン

□□ furan
フラン

□□ pyrimidine
ピリミジン

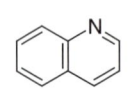

□□ indole
インドール

□□ thiazole
チアゾール

□□ dioxane
ジオキサン

□□ quinoline
キノリン

□□ imidazole
イミダゾール

#41 炭化水素

□□ benzene
ベンゼン
alkene と同じように
[iːn]が語尾

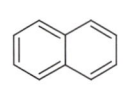

□□ anthracene
アントラセン

□□ toluene
トルエン

□□ o-xylene
o-キシレン

□□ styrene
スチレン

C_{60}
□□ fullerene
フラーレン

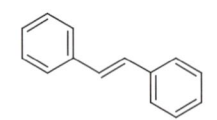

□□ naphthalene
ナフタレン

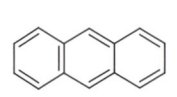

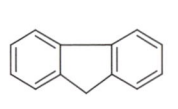

□□ fluorene
フルオレン

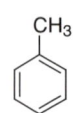

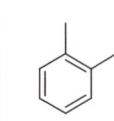

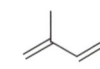

□□ isoprene
イソプレン

□□ *trans*-stilbene
トランス-スチルベン

2.3 化合物の名称

#42 置換炭化水素

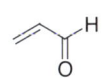

- □□ acrolein
アクロレイン
（発音注意）

- □□ maleic anhydride
無水マレイン酸

- □□ anisole
アニソール

- □□ nitromethane
ニトロメタン

- □□ methacrolein
メタクロレイン

- □□ o-cresol
o-クレゾール

- □□ adipic acid
アジピン酸

- □□ aniline
アニリン

- □□ methacrylic acid
メタクリル酸

- □□ succinic acid
コハク酸

- □□ picric acid
ピクリン酸

- □□ aspirin
アスピリン（商標名）
- □□ acetylsalicylic acid
アセチルサリチル酸

#43 ハロゲン化物，有機リン化合物

- □□ vinyl chloride
塩化ビニル

- □□ methyl bromide
臭化メチル

- □□ deuterated chloroform
重クロロホルム

- □□ triphenylphosphine
トリフェニルホスフィン

- □□ allyl bromide
臭化アリル

- □□ dichloromethane
- □□ methylene chloride
ジクロロメタン

- □□ chloroprene
クロロプレン

- □□ triphenylphosphine sulfide
トリフェニルホスフィンスルフィド

- □□ aryl iodide
ヨウ化アリール

- □□ acetyl chloride
塩化アセチル

#44 無機化合物

HCl(aq)
☐☐ hydrochloric acid
塩酸(塩化水素: hydrogen chloride の水溶液)

H₃PO₄
☐☐ phosphoric acid
リン酸

NaCl
☐☐ sodium chloride
塩化ナトリウム

MgO
☐☐ magnesium oxide
酸化マグネシウム

O₃
☐☐ ozone
オゾン

KCN
☐☐ potassium cyanide
シアン化カリウム

MnO₂
☐☐ manganese dioxide
二酸化マンガン

HNO₃
☐☐ nitric acid
硝酸

NH₃
☐☐ ammonia
アンモニア

SOCl₂
☐☐ thionyl chloride
塩化チオニル

LiH
☐☐ lithium hydride
水素化リチウム

H₂SO₄
☐☐ sulfuric acid
硫酸

NaOH
☐☐ sodium hydroxide
水酸化ナトリウム

NaHCO₃
☐☐ sodium bicarbonate
炭酸水素ナトリウム

H₂O₂
☐☐ hydrogen peroxide
過酸化水素

#45

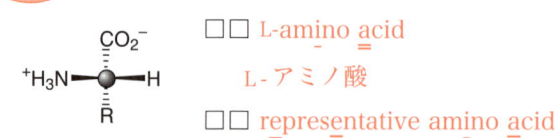

☐☐ L-amino acid
L-アミノ酸

☐☐ representative amino acid

☐☐ alanine
アラニン(Ala)

☐☐ valine
バリン(Val)

☐☐ isoleucine
イソロイシン(Ile)

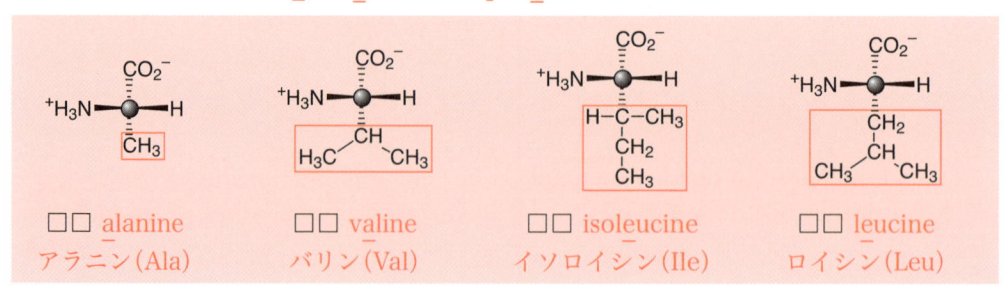

☐☐ leucine
ロイシン(Leu)

□□ methionine
メチオニン(Met)

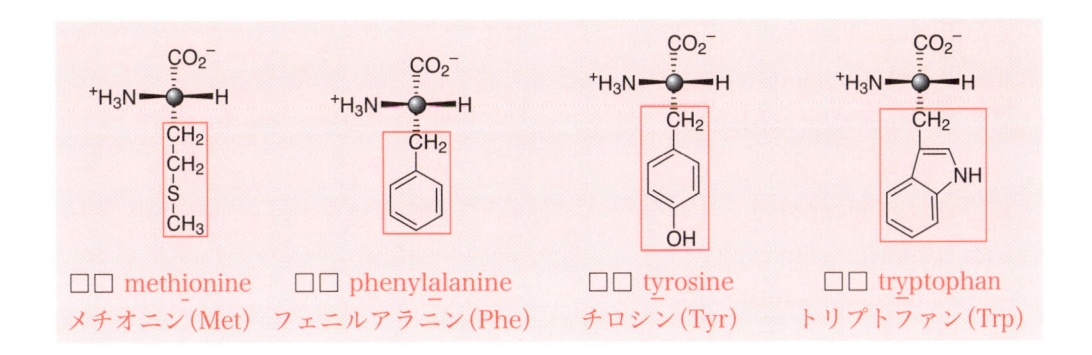

□□ phenylalanine
フェニルアラニン(Phe)

□□ tyrosine
チロシン(Tyr)

□□ tryptophan
トリプトファン(Trp)

2.4 人名反応など

外国人の名前も日本人にとっては聞き取りや発音が難しい．音声ファイルを利用して，自ら発音して覚えよう．

#46

□□ aldol reaction　　　　　　　　　アルドール反応
□□ Arrhenius equation　　　　　　　アレニウス式
□□ Baeyer-Villiger oxidation　　　　バイヤー‐ビリガー酸化
□□ Baldwin rule　　　　　　　　　　ボールドウィン則
□□ Beckmann rearrangement　　　　ベックマン転位
□□ Birch reduction　　　　　　　　バーチ還元
□□ Brønsted acid　　　　　　　　　ブレンステッド酸
□□ Brookhart catalysts　　　　　　ブルックハルト触媒
□□ Claisen condensation　　　　　クライゼン縮合
□□ Claisen rearrangement　　　　　クライゼン転位
□□ Clemmensen reduction　　　　　クレメンゼン還元
□□ Cope rearrangement　　　　　　コープ転位
□□ Cram's rule　　　　　　　　　　クラム則
□□ Darzens reaction　　　　　　　ダルツェンス反応
□□ Dewar-Chatt bonding model　　デュバール‐チャットモデル
□□ Dieckmann condensation　　　　ディークマン縮合
□□ Diels-Alder reaction　　　　　　ディールス‐アルダー反応
□□ Eschenmoser reaction　　　　　エッシェンモーザー反応

- ☐☐ Eyring equation　　　　　アイリングの式
- ☐☐ Felkin-Ahn model　　　　　フェルキン‐アーンモデル
- ☐☐ Friedel-Crafts alkylation　　フリーデル‐クラフツアルキル化
- ☐☐ Grignard reagent　　　　　グリニャール試薬
- ☐☐ Grubbs catalyst　　　　　　グラブス触媒
- ☐☐ Hammet relationship　　　　ハメットの関係
- ☐☐ Hammond postulate　　　　　ハモンドの仮説
- ☐☐ Hartwig-Buchwald coupling　ハートウィック‐バックウォルドカップリング
- ☐☐ Heck reaction　　　　　　　ヘック反応
- ☐☐ Horner-Wadsworth-Emmons reaction　ホーナー‐ワズワース‐エモンズ反応
- ☐☐ Hückel's rule　　　　　　　ヒュッケル則
- ☐☐ Jahn-Teller distortion　　ヤーン‐テラーひずみ
- ☐☐ Jones oxidation　　　　　　ジョーンズ酸化
- ☐☐ Kaminsky catalyst　　　　　カミンスキー触媒
- ☐☐ Karplus equation　　　　　　カープラス式
- ☐☐ Kolbe coupling　　　　　　　コルベカップリング
- ☐☐ Lewis acid　　　　　　　　　ルイス酸
- ☐☐ Lindlar catalyst　　　　　　リンドラー触媒
- ☐☐ Mannich reaction　　　　　　マンニッヒ反応
- ☐☐ Markovnikov rule　　　　　　マルコフニコフ則
- ☐☐ McMurry coupling　　　　　　マクマリーカップリング
- ☐☐ Meerwein-Ponndorf reduction　メーヤワイン‐ポンドルフ還元
- ☐☐ Merrifield method　　　　　メリフィールド法
- ☐☐ Michael addition　　　　　マイケル付加
- ☐☐ Mitsunobu reaction　　　　光延反応
- ☐☐ Monsanto process　　　　　モンサント法
- ☐☐ Newman projection　　　　　ニューマン投影図
- ☐☐ Oppenauer oxidation　　　　オペンナウアー酸化
- ☐☐ Pauli exclusion principle　パウリの排他理論
- ☐☐ Pauson-Khand reaction　　　ポウソン‐カーン反応
- ☐☐ Perkin reaction　　　　　　パーキン反応
- ☐☐ Peterson elimination　　　ピーターソン脱離

☐☐	Reformatsky reaction	リフォーマトスキー反応
☐☐	Robinson cyclization	ロビンソン環化
☐☐	Schrock carbene	シュロックカルベン
☐☐	Sharpless asymmetric epoxidation	シャープレス不斉エポキシ化反応
☐☐	Shiff base	シッフ塩基
☐☐	Simmons-Smith reaction	シモンズ-スミス反応
☐☐	Sonogashira coupling	薗頭カップリング
☐☐	Stille coupling	スチレカップリング
☐☐	Suzuki coupling	鈴木カップリング
☐☐	Swern oxidation	スワン酸化
☐☐	Tamao coupling	玉尾カップリング
☐☐	Tebbe's reagent	テッベ試薬
☐☐	Wacker oxidation	ワッカー酸化
☐☐	Willkinson's catalyst	ウィルキンソン触媒
☐☐	Wittig reaction	ウィッティッヒ反応
☐☐	Wolf-Kishner reduction	ウォルフ-キシュナー還元
☐☐	Wolf rearrangement	ウォルフ転位
☐☐	Woodward-Hoffmann rule	ウッドワード-ホフマン則
☐☐	Wurtz reaction	ウルツ反応
☐☐	Zeise's salt	ツァイズ塩
☐☐	Ziegler-Natta catalyst	チーグラー-ナッタ触媒

2.5　その他の反応名など

#47

☐☐	oxidative addition	酸化的付加
☐☐	reductive elimination	還元的脱離
☐☐	transmetalation	金属交換反応, トランスメタル化
☐☐	β-elimination	β-脱離
☐☐	hapticity	ハプト数
☐☐	back donation	逆供与
☐☐	ligand field theory	配位子場理論

☐☐ spectrochemical series　　　分光化学系列
☐☐ agostic interaction　　　アゴスチック相互作用
☐☐ cone angle　　　円錐角，コーンアングル
☐☐ bite angle　　　配位挟角
☐☐ π–allyl　　　π－アリル
☐☐ σ–allyl　　　σ－アリル
☐☐ metallacycle　　　含金属環化物
☐☐ carbopalladation　　　カーボパラデーション
☐☐ hydroboration　　　ヒドロホウ素化
☐☐ hydrometallation　　　ヒドロメタル化
☐☐ asymmetric amplification　　　不斉増幅（触媒の e.e. より高い e.e. が発現すること）
☐☐ asymmetric multiplication　　　不斉増殖（触媒量の不斉源から生成物を得ること）
☐☐ metathesis　　　メタセシス
☐☐ ring-opening metathesis polymerization（ROMP）　　　環化メタセシス重合

2.6　化学にかかわる者として最低限知っておきたい英単語

2.6.1　学生・研究生活関連の単語

#48

☐☐ term　　　（3学期制の）学期
☐☐ semester　　　（2学期制の）学期
☐☐ credit　　　単位
☐☐ syllabus　　　シラバス
☐☐ grade　　　（米）成績＝（英）mark
☐☐ tuition　　　授業料・授業
☐☐ national university　　　国立大学
☐☐ state university　　　州立大学
☐☐ private university　　　私立大学
☐☐ independent administrative corporation（または agency または institution）　　　独立行政法人

☐☐ machine scored exam　　　マークシート試験
☐☐ national center test for university administration　　　センター試験
☐☐ listening comprehension test　　　ヒヤリング試験　（hearing test は聴覚試験）
☐☐ oral exam/defense　　　口頭試問
☐☐ entrance ceremony　　　入学式
☐☐ graduation ceremony　　　（高校までの）卒業式

☐☐ commencement	（大学の）卒業式	☐☐ graduation thesis	卒業論文
	（commence は「始まる」の意）	☐☐ major in ～	～を専攻する
☐☐ class reunion	同窓会(催し)	☐☐ be located	位置する
☐☐ alumni association	同窓会(組織)	☐☐ compulsory education	義務教育
☐☐ academic year	学術年	☐☐ graduate school	大学院
	（日本では4月～3月の期間をさす）		

2.6.2　学年・職位・学位などの単語

#49

☐☐ freshman	学部1年生	☐☐ master's degree	修士号
☐☐ sophomore	学部2年生	☐☐ doctorate	博士号
☐☐ junior	学部3年生	☐☐ postdoctoral fellow	
☐☐ senior	学部4年生		博士研究員，ポスドク
☐☐ undergraduate	学部学生	☐☐ lecturer	（大学の）講師，講演者
☐☐ grad	大学院生(口語)	☐☐ assistant professor	助教
☐☐ graduate	大学院生(米)，卒業する	☐☐ associate professor	准教授
☐☐ postgraduate	大学院生(英)	☐☐ professor	教授
☐☐ bachelor's degree	学士号	☐☐ dean	学部長
☐☐ Bachelor of Arts (BA)	文学士号	☐☐ president	学長

2.6.3　教科名などの単語

#50

☐☐ organic chemistry	有機化学	☐☐ physics	物理学
☐☐ inorganic chemistry	無機化学	☐☐ biology	生物学
☐☐ physical chemistry	物理化学	☐☐ pharmacy	薬学
☐☐ applied chemistry	応用化学	☐☐ medical science	医学
☐☐ synthetic chemistry	合成化学	☐☐ natural science	理科
☐☐ biochemistry	生化学	☐☐ liberal arts	（一般）教養科目
☐☐ analytical chemistry	分析化学	☐☐ required subject	必修科目
☐☐ material chemistry	材料化学	☐☐ elective (subject)	選択科目
☐☐ chemical engineering	化学工学	☐☐ laboratory (lab)	ラボ，研究室
☐☐ quantum mechanics	量子力学		

2.6.4 実験関連の単語

#51

☐☐ microscope	顕微鏡	☐☐ pipet(te)	ピペット
☐☐ scale	目盛り，天秤	☐☐ reagent bottle	試薬瓶
☐☐ balance	天秤	☐☐ funnel	ろうと
☐☐ syringe	注射器	☐☐ separatory funnel	分液ろうと
☐☐ needle	針	☐☐ stir bar	回転子
☐☐ steel cylinder	ボンベ	☐☐ thermometer	温度計
(独語の Bombe が語源)		☐☐ fume hood	ドラフト
☐☐ tweezers	ピンセット	☐☐ condenser	冷却管
☐☐ spatula	スパチュラ，料理用のへら	☐☐ sample tubing	サンプル管
☐☐ desiccator	デシケーター	☐☐ test tube	試験管
☐☐ three-way stopcock	三方コック	☐☐ test tube rack	試験管立て
☐☐ petri dish	シャーレ	☐☐ rubber stopper	ゴム栓
☐☐ weighing paper	薬包紙	☐☐ magnetic stirrer	スターラー
☐☐ bench	実験台	☐☐ aspirator	アスピレーター
☐☐ two-necked flask	二口フラスコ	☐☐ evaporator	エバポレーター
☐☐ Erlenmeyer flask	三角フラスコ	☐☐ safety glasses	安全眼鏡
☐☐ pear shape flask	梨型フラスコ	☐☐ vacuum pump	真空ポンプ
☐☐ volumetric flask	メスフラスコ	☐☐ ultrasonic cleaner	超音波洗浄器
☐☐ beaker	ビーカー	☐☐ chromatography	クロマトグラフィー
☐☐ buret（米）・burette	ビュレット		

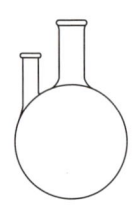

two-necked flask 　　Erlenmeyer flask 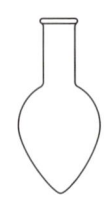　　pear shape flask 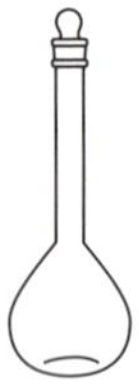　　volumetric flask

Chapter ❸ 化学に頻出する基本構文を覚えよう

Chapter 1 では発音, Chapter 2 では単語について学習した. この Chapter 3 では, 化学英語で重要な構文を中心に学習する. 一般の英語とは異なる意味で使用される語があることに注意しよう. 次の Chapter 4 の例文を学習すると実感できるはずだが, ここにあげた 150 の文例を理解できれば, かなりの化学英語に対応できる. また, これらの構文を覚えれば英作文にもおおいに役立つ(問題集あり. p. v 参照).

3.1 英語の文章を声にだして読むときの注意

本書の最初でオーバーラッピングとシャドーイングの重要性について述べたが, Chapter 3 でも音声ファイルを十分活用し, リスニングとスピーキングによって英語力をつけよう. ここでは, 英語の文章を読むときに注意すべき事柄について以下にまとめた.

① 単語間にポーズを入れない

まず次のページに示した例文1を, 実際に声をだして読んでみよう. はじめのうちは, いくら速く読もうとしても, 「ザ」(間)「ターム」(間)「アブソルート」…と一つ一つの単語が独立してしまう. 駅の列車案内や駐車場の自動支払い計算機などで, 録音した単語と単語を機械的に合成した不自然な日本語をよく耳にするが, 日本人の英語はこれとよく似ている. 「ザタームアブソルート…」のように, 前の単語の最後の音と次の単語の先頭の音をつなげて読んでみよう. 何度も何度も繰り返し読む練習をすると, 意識せずとも自然に音がつながっていくはずだ.

② アクセントをつけてみる

「ザタームアブソルート…」といえるようになったら, 次はアクセントを入れよう.「ザタームアブソルート…」と下線の所を強く読み, 逆に下線のない語は, 今までより弱く読んでみる. そうすることで, かなり英語らしいサウンドになるはずである).

③リズムをつけてみよう

「ザタームアブソルートイーサーリファーズツーイーサーザットハズビーンドライドオーバーメタルックソディアム」と全文を一本調子で読むのではなく,「ザタームアブソルートイーサー（間）リファーズツーイーサー（間）ザットハズビーンドライドオーバーメタルックソディアム」と, 入れるべきところにはしっかり長めの pause を入れよう. また, タームのター, イーサーのイー, リファーのファーなどの長音は, はっきり長めに発音するとリズムをつけやすい.

連続的に読もうとするとアクセントがつかず, 逆にアクセントをつけようとすると断続的な音になりやすい. 何度も繰り返して練習しよう.

④ elision（音節の省略）に気をつけよう

t, d, k, g, p, b などの破裂音が語尾に来ると, 弱く発音されることがある. また, 弱く発音される l は日本人には聞き取りにくい場合がある. たとえば例文 7 の all reagent, available は「オーリエイジェント」,「アベイラボ」となりうる.

3.2　実践トレーニング 構文150

#52 Lesson 1

1. □□ The term absolute ether **refers to** ether that has been dried over metallic sodium.

   無水エーテルとは, 金属ナトリウムを用いて乾燥させた（ジエチル）エーテルのことを指す.

   Point absolute には「絶対の」と「無水の」の二つの意味がある. absolute zero temperature とは絶対零度のこと.

   Point ether は, 特定の物質である「ジエチルエーテル」そのものを指すときと,「エーテル類」という種類を指すときがある. 特定の物質そのものを指すときは, その単語は「物質名詞」で「不可算名詞」となり, 不定冠詞 a がついたり複数形になることはない. 例文ではジエチルエーテルそのものを指し示している.

   Point 関係代名詞（主格・目的格）には,「〜するところの物・事」を意味する制限用法（コンマなしで用いる）と,「そしてそれ（ら）は〜」の意の非制限用法がある. 制限用法

は that も which も使用可．非制限用法では that は通常不可であり，コンマをつけて 〜, which とする．

> **Jeff's advice**: 私の場合，制限用法では which ではなく that を好んで使う．

2. □□ This result **indicates that** the catalyst is not necessary.

   この結果は，その触媒が必要ではないことを意味している．

Point 「意味する」は一般には mean がまず思い浮かぶが，論文などで実験事実に基づいて「〜を意味する」という場合は，indicate, show, demonstrate が用いられることが多い．

3. □□ The step with the highest activation energy **is called** the rate-limiting step.

   最も活性化エネルギーの高い段階は，律速段階と呼ばれる．

Point この文章を能動態にもどすと，We call the step with the highest activation energy the rate-limiting step. となる．すなわち，例文で 〜 is called as the rate 〜 と as を入れてはいけないことに注意．

Point rate-limiting は化学量論反応にも触媒反応にも用いられるが，rate-determining はサイクル中のすべてのステップが同じ速度で進行する触媒反応に用いるには適当でない．

> **Jeff's advice**: 同じ訳語の単語でも指し示す内容が違う．

4. □□ In stark contrast, the attempted reaction **provided compound A**.

   きわめて対照的に，試行実験は**化合物 A を生じた**．

Point In stark contrast は導入句である．一般に導入句のあとにはコンマを入れて，主文との区別をわかりやすくする．

5. □□ Formic acid **undergoes decomposition** when it is heated.

   ギ酸は加熱すると**分解する**．

40　● Chapter 3　化学に頻出する基本構文を覚えよう

$$HCO_2H \xrightarrow{加熱} H_2 + CO_2$$

Point undergo a similar reaction（類似の反応を起こす）など，undergo ～ で「（反応・変化などを）起こす」の意味で多用される表現．

Point formic acid は特定の化学物質名であるため「物質名詞」で，不可算名詞扱い．

6. ☐☐ The reaction **takes place** through a complex mechanism.　｜　その反応は，複雑なメカニズムを経て起こる．

Point take place は occur と同様，「（反応が）起こる」という意味で最もよく使われる表現．和英辞典で「起こる」を引くと happen がまずヒットするが，「偶然に起こる」というニュアンスを含むので「反応が起こる」の意で使われることはまれ．

Point the complex mechanism shown in Scheme 5 というようなある特定のメカニズムを指すときは定冠詞 the をつける．

> ここは特定のものを指さないので不定冠詞でよい．
> Jeff's advice

7. ☐☐ All reagents **are commercially available**.　｜　すべての試薬は市販されている．

Point be commercially available は，覚えておきたい英語特有の表現の一つ．

8. ☐☐ The reaction mixture was stirred **at −78 ℃ for 30 min** and then **allowed to warm to** ambient temperature.　｜　反応混合物を −78 ℃ で 30 分撹拌し，その後，室温まで昇温した．

Point ℃は degrees Celsius または degrees centigrade と発音する．degrees と複数形の s をつけることに注意．

Point 主語が共通であり，and then の前にコンマは不要．

3.2 実践トレーニング 構文150

9. ☐☐ Compound **A** reacted with compound **B** under the same reaction conditions.

化合物 **A** と化合物 **B** とは，それとまったく同じ条件下で反応した．

Point compound **A** は固有名詞扱い．固有名詞は無冠詞でよい．例文23参照．
Point the same ～とするのに対し，the similar ～は不可．「類似の」は a similar ～あるいは similar ～s とする．
Point **A** reacts with **B** の基本構文が使われていることに着目(6.2.2参照)．

10. ☐☐ The SO group in dimethyl sulfoxide is highly polar. In other words, the dipole moment is very large.

ジメチルスルホキシドの SO 基は大きく分極している．いい換えると，双極子モーメントが非常に大きい．

Point dimethyl sulfoxide は特定の物質そのものを指すので冠詞は不要（例文1参照）．ただし，in dimethyl sulfoxide のあとに molecule をつけると，in the dimethyl sulfoxide molecule と定冠詞 the が必要になる．これは，dimethyl sulfoxide が形容詞の働きをして molecule が限定されるからである

> 冠詞が必要かどうかに注意が必要である．
> Jeff's advice

Point 「数量が大きい」は，big ではなく large が通常使用される．

Keywords

☐☐ term	用語	☐☐ step	段階
☐☐ absolute	無水の，絶対の	☐☐ activation energy	活性化エネルギー
☐☐ ether	エーテル，ジエチルエーテル	☐☐ rate-limiting step	律速段階
☐☐ refer to ～	～について言及する	☐☐ in stark contrast	きわめて対照的に
☐☐ dried	乾燥させた	☐☐ attempt	試みる
☐☐ metallic	金属の	☐☐ reaction	反応
☐☐ sodium	ナトリウム	☐☐ provide	生じる
☐☐ result	結果	☐☐ compound	化合物
☐☐ indicate	示唆する	☐☐ formic acid	ギ酸
☐☐ catalyst	触媒	☐☐ undergo	（反応などを）起こす
☐☐ necessary	必要な	☐☐ decomposition	分解

Chapter 3 化学に頻出する基本構文を覚えよう

☐☐ heat	加熱する	☐☐ compound	化合物
☐☐ take place	(反応が)起こる	☐☐ react with 〜	〜と反応する
☐☐ complex	複雑な	☐☐ condition	条件
☐☐ mechanism	メカニズム	☐☐ group	基
☐☐ reagent	試薬	☐☐ dimethyl sulfoxide	ジメチルスルホキシド
☐☐ commercially	商業的に		
☐☐ available	利用できる	☐☐ molecule	分子
☐☐ mixture	混合物	☐☐ polar	極性をもつ
☐☐ stir (発音注意)	撹拌する	☐☐ in other words	言い換えると
☐☐ be allowed to 〜	〜する	☐☐ dipole	双極子
☐☐ warm	暖める	☐☐ moment	モーメント
☐☐ ambient temperature	室温		

#53 **Lesson 2**

11. ☐☐ **The reaction rate is** high.　　|　反応速度は**大きい**．

Point 速度は high あるいは low で表現する．big, small は不可．

12. ☐☐ The target molecule **was obtained quantitatively**.　　|　標的分子が**定量的に得られた**．

Point quantitative(定量的な)に対応する「定性的な」は qualitative．

13. ☐☐ The color **remained unchanged** during the course of the reaction.　　|　反応の途中，溶液の色に**変化はなかった**．

Point remain unaffected (影響を受けない), remain unsolved (未解決なまま) など，remain 〜で「〜のままである」の意味で多用される．

14. ☐☐ **Addition of A into** the reaction mixture of B and C **resulted in** an exothermic reaction. | BとCの混合液にAを加えると発熱反応が起きた．

Point result in failure（失敗に終わる）など，result in ～ で「～をもたらす，～の結果となる」の意味で多用される．

15. ☐☐ The present procedure provides a practical **method for** the preparation of these samples. | この操作は，これらの試料を調製するための実用的な**方法**である．

Point 「～の方法」は，method for ～ あるいは method of ～．

16. ☐☐ Then the aqueous solution **is saturated with** sodium chloride. | 次に，水溶液を NaCl で**飽和させる**．

Point この文を能動態にすると，Then we saturated the aqueous solution with sodium chloride. となるが，化学論文（とくに実験項）では人称代名詞を主語に用いる表現はまれである．

17. ☐☐ The flask **is filled with** about 3 mL of liquid ammonia. | そのフラスコを約 3 mL の液体アンモニアで**満たす**．

Point 3 mL は，milliliters と s を付ける．リットルは大文字で書くのが化学では慣例．

18. ☐☐ The precipitated catalyst was **filtered off**. | 析出した触媒は，**ろ過により取り除いた**．

19. ☐☐ We need to **carry out the reaction** to get information about the reaction mechanism. | 反応機構についての情報を得るために，その反応を実行する必要がある．

Chapter 3 化学に頻出する基本構文を覚えよう

Point information about は「インフォーメ ショナ バゥ」と読まれる可能性がある．このように語尾の子音と次の語の頭母音とを続けて読むことを liaison（音の連結）と呼ぶ．

Point route は本来「道」を意味するので，reaction route よりも reaction pathway あるいは reaction mechanism のほうが適している．

> 化学的に明確な意味を表す単語を使おう．
> Jeff's advice

20. ☐☐ An authentic sample **was furnished by** a known procedure.
標準品は，既知の方法により合成した．

Point 「（化学反応により）得られた」を表す動詞としては，give, produce, provide, afford, yield, furnish, generate, form などが用いられる．6.2.2 を参照．

Keywords

☐☐ rate	（反応）速度	☐☐ preparation	調製
☐☐ target	標的	☐☐ sample	試料
☐☐ molecule	分子	☐☐ aqueous	水溶性の
☐☐ obtain	得る	☐☐ saturate	飽和させる
☐☐ quantitatively	定量的に	☐☐ solution	溶液
☐☐ color	色	☐☐ flask	フラスコ
☐☐ remained unchanged	変化しないままである	☐☐ be filled with ～	～で満たす
		☐☐ liquid	液体
☐☐ course	経過	☐☐ ammonia	アンモニア
☐☐ addition	付加	☐☐ precipitated	沈殿した
☐☐ result in ～	～という結果になる	☐☐ be filtered off	ろ過する
☐☐ exothermic	発熱を伴う，発熱の	☐☐ carry out	（反応を）実行する，行う
☐☐ procedure	手順，手続き	☐☐ mechanism	機構，メカニズム
☐☐ practical	実用的な	☐☐ authentic	標準の，真正の
☐☐ method for ～	～の方法	☐☐ furnish	（生成物を）与える

#54 Lesson 3

21. ☐☐ Drastic measures **are needed to** improve the efficiency.

    効率改善のためには，思い切った手だてが必要である．

22. ☐☐ We have **encountered irreproducibility of the results**.

    われわれの**結果**には，**再現性**がなかった．

23. ☐☐ The reaction route we have envisioned **is shown in** Scheme 1.

    われわれが思い描く反応経路を図1に示した．

Point Scheme 1 などは固有名詞として働くため，冠詞は必要ない（例文9参照）．連続した反応式，反応機構などは Scheme，X線の ORTEP やグラフなどの図は Figure で示されることが多い．

24. ☐☐ **That is true to the extent that** only one isomer is produced.

    ただ一つの異性体のみしか生成しないという点においては，それは正しい．

25. ☐☐ The component **remained intact**（アクセント位置注意）through the treatment.

    その構成部分は，その処理では**完全に保**たれていた．

Point through the treatment は，after the treatment や during the treatment ともできる．

> by the treatment は自然な英語ではない．
> Jeff's advice

26. ☐☐ Nitrogen gas **was evolved vigorously** when a solution containing the reagent was heated.

    その試薬を含む溶液を加熱すると，窒素ガスが**激しく発生した**．

27. ☐☐ The extracted organic layer **was distilled under** reduced pressure.

    抽出した有機層を減圧下で蒸留した．

Point 「加圧下で」は under pressure．

28. In general, the *m*-methoxy group is an electron-withdrawing group, while the *p*-methoxy group is an electron-donating group.

一般に，*m*-メトキシは電子求引基であるが，*p*-メトキシは電子供与基である．

> MeO 基や Cl など，孤立電子対をもつ置換基は，σ結合を通しての電子の授受(誘起効果)および，その電子対の供与(共鳴効果)を両方加味する必要がある．MeO 基がベンゼン環の *p*-位にあるときは，共鳴効果が誘起効果を上回り，電子供与基となるのに対し，*m*-位の場合，共鳴効果が有効に働かず電子求引基となる．

Point while と although はともに「〜であるが」だが，前者は並列的なニュアンスが強い．

29. Esterification is a reversible reaction.

エステル化は可逆反応である．

30. The nitro group in nitrobenzene is conjugated with the phenyl ring.

ニトロベンゼンのニトロ基は，フェニル基(環)と共役している．

Keywords

□□	measure	方法，測る	□□ treatment	処理
□□	improve	改良する	□□ nitrogen	窒素
□□	efficiency	効率の良さ	□□ evolve	(気体，光などを)発生する
□□	encounter	でくわす	□□ vigorously	激しく
□□	irreproducibility	再現性のなさ	□□ contain	含む
□□	envision	心に描く	□□ extract	抽出する
□□	scheme	図式，図解	□□ organic	有機の
□□	extent	範囲，限度	□□ layer	層
□□	isomer	異性体	□□ distill	蒸留する
□□	produce	与える	□□ under reduced pressure	減圧下で
□□	component	構成要素	□□ in general	一般的に
□□	intact	完全なままで	□□ methoxy	メトキシ

☐☐ electron-withdrawing	電子求引性の	☐☐ nitrobenzene	ニトロベンゼン
☐☐ electron-donating	電子供与性の	☐☐ conjugate	共役させる
☐☐ esterification	エステル化	☐☐ phenyl	フェニル
☐☐ reversible	可逆の	☐☐ ring	環
☐☐ nitro	ニトロ		

#55 **Lesson 4**

31. ☐☐ Delocalization of electrons **resulted in** the extra binding energy.

電子の非局在化の結果，さらなる結合エネルギーを生じた．

32. ☐☐ Nucleophilic addition of Grignard reagents to aldehydes **is widely employed for** synthesis of alcohols.

グリニャール試薬のアルデヒドへの求核付加は，アルコールの合成**に広く使用されている**．

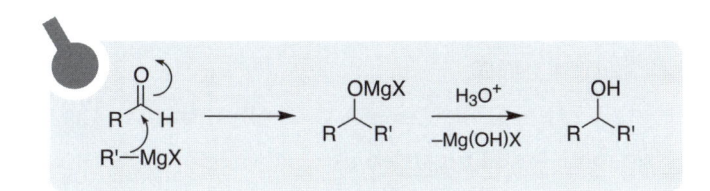

Point 「使用する」の意味では，use と employ が最もよく使われる．一方，utilize は少し堅い表現であり，utilize A for B（動名詞）で「A を B のために使う」というように用いる．

Point 例文1の ether は，特定の化合物を指し，物質名詞であり不可算名詞であった．一方，この例文の Grignard reagent, aldehyde, alcohol はすべてある特定の化学物質を指しているのではなく，いろいろな種類がある化合物群を指しているので可算名詞扱いで，すべて 〜s となっている．

33. ☐☐ **The reaction proceeded** smoothly even **in the absence of** the catalyst.

その反応は，触媒がなくても円滑に**進行した**．

Point smoothly の「th」の発音は [z] ではなく [ð] である．p.9 を参照．

34. ☐☐ The base **abstracts** a proton **from** the ketone **to give** anion **C**.

その塩基は，ケトンからプロトンを引き抜き，アニオン **C** が生成する．

Point　「to ～不定詞」の結果用法は，化学論文では多用される．

35. ☐☐ The ethyl group can **be considered as** a moiety of ethyl acetate.

エチル基は，酢酸エチルの一部分と見なすことができる．

Point　例文 1 の ether と同様，ethyl acetate は一つの化合物名で物質名詞であり，a が付いたり複数形になったりしない．

36. ☐☐ It is necessary to maintain strictly anhydrous conditions to **avoid decomposing** the starting materials.

出発物質の分解を防ぐため，厳密な無水条件を保つ必要がある．

Point　avoid は動名詞を目的語にとる代表的な動詞．そのほか finish, consider, postpone, suggest も同様．

37. ☐☐ This classic route **is** still **regarded as** the best method for the preparation of the compound.

この古典的なルートは，その化合物合成の最良の方法であると今でも見なされている．

38. ☐☐ The reaction **was quenched by** pouring 250 mL of saturated（アクセント位置注意）aqueous sodium bicarbonate into the reaction mixture.

その反応混合物のなかに 250 mL の炭酸水素ナトリウムの飽和水溶液を加えることで，反応をクエンチ（停止）させた．

Point　反応後，残った活性な試薬を含む溶液に水などを加えて反応を停止する操作を，quench という．

3.2 実践トレーニング 構文150

39. ☐☐ Hydrogen bonding is **no longer** possible under such harsh conditions.

そのような過酷な条件では，もはや水素結合は可能ではない．

Point hydrogen bonding は「水素結合」という概念を示し，通常不可算名詞扱い．一方，hydrogen bond は結合そのものを指し，可算名詞扱いし，冠詞が必要となる．

冠詞の有無に注意しよう．
Jeff's advice

40. ☐☐ The reaction of **A** with **B** furnished **C** and **D** in 20% and 30% yields, **respectively**.

AとBとの反応で，CとDがそれぞれ20%，30%の収率で得られた．

Point 「おのおの」というときの決まり表現．語尾に ,(カンマ)＋ respectively.

Keywords

☐☐ delocalization	非局在化
☐☐ extra	余分の
☐☐ binding	結合の
☐☐ nucleophilic	求核的な
☐☐ Grignard reagent	グリニャール試薬
☐☐ aldehyde	アルデヒド
☐☐ employ = use	使用する
☐☐ synthesis	合成
☐☐ alcohol	アルコール
☐☐ proceed	進行する
☐☐ smoothly	円滑に，なめらかに
☐☐ in the absence of ～	～の非存在下で
☐☐ catalyst	触媒
☐☐ base	塩基
☐☐ abstract	引き抜く
☐☐ proton	プロトン，陽子
☐☐ ketone	ケトン
☐☐ anion	アニオン
☐☐ ethyl	エチル
☐☐ be considered as ～	～と見なせる
☐☐ moiety	一部分
☐☐ acetate	アセテート
☐☐ maintain	保持する，保つ
☐☐ strictly	厳密に
☐☐ anhydrous	無水の
☐☐ avoid ～ing	～するのを防ぐ
☐☐ decompose	分解する
☐☐ starting material	出発物質
☐☐ classic	古典的な
☐☐ route	ルート，経路
☐☐ be regarded as ～	～と見なされる
☐☐ method	方法

Chapter 3 化学に頻出する基本構文を覚えよう

- □□ be quenched by 〜
  〜でクエンチする，〜で停止させる
- □□ pour（発音注意） 注ぐ
- □□ saturated 飽和した
- □□ sodium bicarbonate 炭酸水素ナトリウム
- □□ hydrogen bonding 水素結合
- □□ possible 可能な
- □□ harsh = severe （条件が）過酷な，厳しい
- □□ yield 収率
- □□ respectively おのおの

#56 Lesson 5

41. □□ **As for the mechanism,** an S_N2-type route has already been established.

    メカニズムに関しては，S_N2型の経路であることがすでに確立されている．

Point 辞書を見ると，as for 〜も as to 〜も「〜についていえば」とあるが，文頭では，as for 〜のほうがよく使われる．

Point S_N2 は母音で始まるので不定冠詞は a ではなく an．

Point 頻度を表す副詞や，already, still のような状況を表す副詞は，助動詞（be 動詞を含む）の直後というのが一般的なルールである．

> has already been のほうが has been already より自然である．
> — Jeff's advice

42. □□ **No reaction occurred** without the additives.

    添加剤なしではまったく反応しなかった．

43. □□ **Many reagents can be reduced prior**（発音注意）**to** the reaction with the chemicals.

    多くの試薬は，その化学薬品と反応する前に還元される可能性がある．

Point reduce は[ridu:s]と[ridju:s]と読まれる場合があることに注意．

44. □□ **The leaving group was expelled by** the nucleophile.

    その脱離基は，求核剤によって追いだされた．

3.2 実践トレーニング 構文150

45. ☐☐ The mechanism of the electron transfer from **A** to **B remains elusive**.
A から **B** への電子移動のメカニズムは，まだよく理解されていない．

46. ☐☐ The reaction **is triggered by** a single electron transfer.
その反応は，1電子移動によって引き起こされる．

47. ☐☐ Functional groups such as Cl and carbonyl groups **tolerate** the present reaction.
塩素やカルボニル基などの官能基は，この反応には耐性がある．

48. ☐☐ Optically active sulfoxides **are useful** chiral auxiliaries **for** asymmetric syntheses.
光学活性スルホキシドは，不斉合成の有用なキラル補助基である．

> S上に孤立電子対があるためスルホキシドは不斉である

Point sulfoxide は化合物の「種類」を表し，可算名詞である．例文 1, 32, 35 参照．

49. ☐☐ The filtration should be carried out **as rapidly as possible**.
できるだけ迅速にろ過する必要がある．

50. ☐☐ Bromine **is subject to substitution by** other functional groups.
臭素は，ほかの官能基によって置換されやすい．

Keywords

☐☐ as for 〜 〔通例文頭で〕〜についていえば	☐☐ occur （反応が）起こる
☐☐ S$_N$2　2分子求核置換反応	☐☐ additive　添加剤
☐☐ establish　確立する	☐☐ reduce　還元する
	☐☐ prior to 〜　〜より前に

Chapter 3 化学に頻出する基本構文を覚えよう

☐☐ chemical	化学薬品	☐☐ optically active	光学活性な
☐☐ leaving group	脱離基	☐☐ sulfoxide	スルホキシド
☐☐ expel	追いだす	☐☐ chiral	キラルな
☐☐ transfer	転移, 移動	☐☐ auxiliary	補助するもの
☐☐ elusive	理解しにくい	☐☐ asymmetric	不斉の
☐☐ trigger	引き起こす	☐☐ filtration	ろ過
☐☐ functional group	官能基	☐☐ bromine	臭素
☐☐ carbonyl group	カルボニル基	☐☐ be subject to ~	~を受けやすい
☐☐ tolerate	耐性がある	☐☐ substitution	置換

#57 Lesson 6

51. ☐☐ He was able to operate all the machines for **pursuing his research**.

彼は, 研究を遂行するためのすべての機械を操作することができた.

Point all machines でも文法的には誤りでないが, all the machines としないと,「彼の研究を遂行するための機械」というニュアンスがでない.

Point アメリカ英語では research と発音することも多い.

定冠詞 the の使い方に慣れよう. — Jeff's advice

52. ☐☐ All attempts to separate the stereoisomers **were unsuccessful**.

立体異性体を分離するすべての試みは不成功だった.

53. ☐☐ The special apparatus **permitted the observation** of spectral changes *in situ*.

その特殊な装置のおかげで, 系中でのスペクトル変化の観測が可能になった.

Point apparatus (アクセント位置注意) は,「(一式となった) 装置」. facility は「設備」. gadget は「(ちょっとした) 装置・器具」.

Point *in situ* はラテン語起源で「系中」の意味. 英語以外の語に由来する語はイタリック体で用いられることが多い.

54. □□ **The** p**u**rity of th**i**s c**o**mpound **was confirmed by** g**a**s chromat**o**graphy（アクセント位置注意）. | この化合物の純度は，ガスクロマトグラフィーで**確かめられた**.

Point g**a**s chr**o**matograph は g**a**s chromat**o**graphy（分析）を行うための装置であり，純度は chromat**o**graphy により確かめる.

> 英語では「機器で確かめる」ではなく「機器による分析で確かめる」と表す.
> Jeff's advice

55. □□ The p**e**ak **is ass**i**gnable to** the π-π* tr**a**nsition. | そのピークは π-π* 遷移に**帰属できる**.

56. □□ **It was ant**i**cipated that** the s**y**stem would be **u**seful for more **a**ccurate an**a**lysis. | そのシステムは，より正確な分析に役立つものと**考えられていた**.

Point an**a**lysis は，使い方により可算名詞としても不可算名詞としても使われる．ここでは，不可算名詞扱いされており，分析を漠然と抽象的にとらえている．ここで an**a**lyses とすると，可算名詞扱いとなり，具体的な分析の行為やその結果を指す．

57. □□ The strong p**e**ak around 1730 cm^{-1} **is due to** the C=O stretching vibr**a**tion of the **e**ster. | 1730 cm^{-1} 付近の強いピークは，そのエステルの C=O 伸縮振動によるものである．

Point 1730 は **o**ne thousand seven h**u**ndred and th**i**rty でもよいが，口語では s**e**venteen h**u**ndred and th**i**rty と読まれることが多い．

> ネイティブの多くは後者で読む．
> Jeff's advice

Point is を以下のように変えると可能性が徐々に低下する．
will be ＞ would be ＞ may/might be ＞ could be

58. □□ The spectra were **a**nalyzed **on the b**a**sis of** method **A**. | 方法 **A** に基づいて，そのスペクトルを分析した．

Chapter 3 化学に頻出する基本構文を覚えよう

59. ☐☐ The deep blue color of the starch-iodine complex **serves as** a very sensitive test for iodine.

ヨウ素−デンプン錯体の深い青色は，非常に感度のよいヨウ素の検出法**として役立つ**．

60. ☐☐ That method may be used to **identify** the products.

あの方法は，生成物を**同定する**のに使えるかもしれない．

Keywords

☐☐	operate	操作する	☐☐ transition	遷移
☐☐	machine	機械	☐☐ anticipate	期待する
☐☐	pursue	（実験などを）遂行する	☐☐ system	システム，系
☐☐	research	研究，調査	☐☐ accurate	正確な
☐☐	attempt	試み	☐☐ analysis	分析
☐☐	separate	分離する	☐☐ be due to 〜	〜による，〜のおかげである
☐☐	stereoisomer	立体異性体	☐☐ stretching vibration	伸縮振動
☐☐	unsuccessful	〜に失敗した	☐☐ ester	エステル
☐☐	special	特殊な	☐☐ spectrum（複数形は spectra）	スペクトル
☐☐	apparatus	（ちょっとした）装置	☐☐ analyze	分析する
☐☐	permit	可能にする	☐☐ on the basis of 〜	〜に基づく
☐☐	observation	観測	☐☐ deep	深い
☐☐	spectral	スペクトルの	☐☐ starch	デンプン
☐☐	change	変化	☐☐ iodine	ヨウ素
☐☐	in situ	系中での	☐☐ complex	錯体
☐☐	purity	純度	☐☐ serve as 〜	〜として役立つ
☐☐	confirm	確かめる	☐☐ sensitive	感度の高い
☐☐	chromatography	クロマトグラフィー	☐☐ test	検査，テスト
☐☐	peak	ピーク	☐☐ identify	同定する
☐☐	be assignable to 〜	〜に帰属できる		
☐☐	π-π^*	π軌道からπ^*軌道への		

#58 Lesson 7

61. ☐☐ He **insisted that** the compound (should) be analyzed in the first place.

彼はまず，その化合物の分析が先だと主張した．

Point 要求・主張などを表す文章中の should は省略可．

62. ☐☐ The changes in the structure **brought about** the red-shift in the spectrum.

構造変化は，スペクトルのレッドシフトをもたらした．

> 可視光吸収スペクトルで吸収波長が長波長側（より赤に近い方向）に移動することを red-shift という．逆に短波長側（より紫に近い方向）の移動することを blue-shift（ブルーシフト）という．

Point bring about activation（活性化をもたらす）など，「bring about + 変化を表す語」は「〜をもたらす」の意味で多用される．

63. ☐☐ The residual solid **was recrystallized from** the ethanol solution.

残った固体は，エタノール溶液から再結晶した．

Point residual は「残った，残余の」という意味．関連する語として resultant（結果として生じた）もセットで覚えておこう．これらの形容詞がつくと名詞の意味が限定されるので，通常は定冠詞 the がつくことにも注意．

64. ☐☐ Twice four **is equal to** eight.

4 の 2 倍は 8 である．

65. ☐☐ **The combination of** eqs 1 and 2 leads to eq 7.

式 1 と式 2 を組み合わせると式 7 が導かれる．

Point eq は equation の略．Eq や Eq. とも略される．化学雑誌の種類によって式・図・表などの書き方のフォーマットが統一されていることが多い．記号に添えて示す

単語は固有名詞扱いとなり無冠詞(例文9, 23を参照).

66. □□ The partial pressure of oxygen **affects** the position of equilibria.

   酸素の分圧が，平衡点に影響をおよぼす．

   Point affect は「悪影響」の意味を含む場合がある．一方，effect は「効果」という意味の名詞で使われることが多い．
   Point equilibria は equilibrium の複数形．

67. □□ The pressure of gas **is (directly) proportional to** the absolute temperature at constant volume.

   気体の圧力は，一定体積の下では絶対温度に(正)比例する．

   Point 「反比例する」は be inversely proportional to ～

68. □□ An ideal gas **obeys the following equation:** $pV = nRT$, where p is the pressure when n moles of a gas occupy a volume V at temperature T (measured in Kelvin).

   理想気体は，次の式に従う：すなわち $pV = nRT$ である．ここで p は，n モルの気体が（Kelvin の単位で測定した）温度 T において体積 V を占めるときの圧力である．

   Point ,(コンマ)→ ;(セミコロン)→ :(コロン)→ .(ピリオド)の順に区切り（分離性・独立性）が強くなる．:(コロン)は，以下に具体例や言い換えなどが示される．コロンはこのほか，論文の副題(subtitle)を示すときにも使われる．また;(セミコロン)は，～(文); however, ～(文)という形で，however などの接続副詞で文を接続詞のようにつなぐときによく用いる．
   Point following も次の名詞を限定する働きがあり，定冠詞 the がつく．

69. □□ **The product of** pressure **and** volume of a gas is constant at a fixed temperature.

   気体の圧と体積の積は，一定の温度下では一定である．

Point product は，化学では「生成物」を意味するが，数学では「積」．また，一般には「製品，生産物」の意．

70. □□ The outcome is caused not by thermodynamic control but by kinetic control.

その結果は，熱力学的支配でなく速度論的支配によるものである．

> ある反応において，「活性化エネルギー」は「反応速度」と，「反応系の相対的安定性（$\Delta G°$：自由エネルギー差）」は「熱力学」と相関があることに注意．つまり，$\Delta G^{\ddagger}$ は速度定数 k と，$\Delta G°$ は平衡定数 K と関係づけられる．

Keywords

□□ insist	主張する	
□□ bring about	もたらす	
□□ red-shift	レッドシフト	
□□ residual	残りの	
□□ solid	固体	
□□ recrystallize	再結晶させる	
□□ be equal to ～	～に等しい	
□□ combination	結合，組合せ	
□□ eq 1 (Eq. 1 や eq. 1 とも書く)	式1 (equation 1 の略)	
□□ lead	導く	
□□ partial pressure	分圧	
□□ oxygen	酸素	
□□ affect	影響する	
□□ position	位置	
□□ equilibrium（複数形は equilibria）	平衡	
□□ is proportional to ～	～に比例する	
□□ constant	一定の	
□□ volume	体積	
□□ ideal gas	理想気体	
□□ obey	従う	
□□ following	次の	
□□ pressure	圧力	
□□ mole	モル	
□□ occupy	占有する	
□□ Kelvin	ケルビン（絶対温度の基本単位）	
□□ product	積	
□□ fixed	固定した，一定の	
□□ outcome	結果	
□□ cause	（結果として）引き起こす	
□□ thermodynamic	熱力学の	
□□ control	支配	
□□ kinetic	速度論の	

#59 Lesson 8

71. ☐☐ **The** single bond in 1,3-butadiene **is not equivalent to** the corresponding one in *n*-butane **in terms of** bond order.

1,3-ブタジエンの単結合は，結合次数の点からいえば *n*-ブタンの単結合と同等ではない．

> **Point** 1,3-butadiene も *n*-butane も物質名詞で，冠詞がついたり複数形になったりしない．

72. ☐☐ **One should notice that** the activation energy of this reaction is, **if anything**, negligibly small.

この反応の活性化エネルギーは，たとえあるにしても無視できるほど小さいことに留意せよ．

> **Point** 論文中では，ought to や had better が使用されることはまれ．

73. ☐☐ A dissolved substance **raises** the boiling point and **lowers** the freezing point of a solvent.

溶解した物質は，溶媒の沸点を上昇させ凝固点を下げる．

74. ☐☐ **We shall be concerned with** the protium whose spin number *I* is 1/2.

スピン数 *I* が 1/2 のプロチウムについて考えてみよう．

> **Point** hydrogen は，^{1}H, ^{2}D, ^{3}T の総称で，^{1}H は protium（2.2 参照）．

75. ☐☐ A square **is defined as** a rectangle with four equal sides.

正方形は，四つの等しい辺をもつ長方形として定義できる．

76. ☐☐ The aromatic rings are **neither** coplanar **nor** perpendicular to one another.
　＝　The aromatic rings are **not** coplanar **or** perpendicular to one another.

それらの芳香環は，互いに共平面にも垂直の関係にもない．

Point one another は代名詞であり，副詞ではないので to が必要．

77. ☐☐ The Ph ring **is adjacent to** the carbonyl group.

そのフェニル基(ベンゼン環)は，カルボニル基に**隣接している**．(Ph = C_6H_5-)

78. ☐☐ **The greater** the initial (アクセント位置注意) concentration, **the shorter** the time required to reduce it by one-half.

初期濃度が**高くなるほど**，濃度が半分に低下するのに要する時間は**短い**．

Point 「The 比較級 〜, the 比較級 〜」で「〜であるほどより〜である」．

> この構文で動詞が be 動詞の場合，通常動詞は省略される．
> — Jeff's advice

79. ☐☐ Kinetic studies of these reactions **are actively in progress**.

これらの反応の速度論的な研究は，精力的に進行中である．

Point kinetic studies に定冠詞 the をつけると，「すでに議論されている特定の研究」を指すことになる．

> 一般的な概念を示したいときは the をつけない．
> — Jeff's advice

80. ☐☐ The starting material **was consumed** within 10 min.

出発物質は，10 分以内に**消費された**．

Keywords

☐☐ single bond	単結合	☐☐ activation energy	活性化エネルギー
☐☐ 1,3-butadiene	1,3-ブタジエン		
☐☐ is equivalent to 〜	〜に相当する	☐☐ if anything	あるにしても
☐☐ corresponding	対応する	☐☐ negligibly	無視してよいほど
☐☐ n-butane = normal butane	ノルマルブタン	☐☐ dissolved	溶解した
		☐☐ substance	物質
☐☐ in terms of 〜	〜の点で	☐☐ raise	上げる
☐☐ bond order	結合次数	☐☐ boiling point	沸点
☐☐ notice	注意する	☐☐ lower	下げる

Chapter 3 化学に頻出する基本構文を覚えよう

- □□ freezing point　凝固点
- □□ be concerned with ～　～を扱う，～に関心をもつ
- □□ protium　プロチウム（質量数1の水素）
- □□ spin　スピン（ここでは核スピンのこと）
- □□ square　正方形
- □□ be defined as ～　～として定義する
- □□ rectangle　長方形
- □□ equal　等しい
- □□ side　辺
- □□ aromatic　芳香族の
- □□ coplanar　共平面の，同一平面上の
- □□ perpendicular　垂直の，直角をなす
- □□ be adjacent to ～　～に近接している
- □□ carbonyl group　カルボニル基
- □□ initial　初期の
- □□ require　必要とする
- □□ actively　積極的に
- □□ in progress　進行中で
- □□ consume　消費する

#60 Lesson 9

81. □□ That reaction **is out of control**.　あの反応は，制御不能である．

Point
out of order　故障中で　　out of the center　中心からずれて
out of phase　位相がずれて　out of the plane　面外の

82. □□ The conductivity **was enhanced** dramatically when the conjugation was extended.　伝導度は，共役が拡張されると劇的に高まった．

83. □□ The reaction **was significantly accelerated by** the addition of $CuCl_2$.　その反応は，塩化銅を添加することで著しく促進された．

Point $CuCl_2$ 塩化第二銅／塩化銅 (II) は，cupric chloride あるいは copper dichloride. 一方，CuCl 塩化第一銅／塩化銅 (I) は，cuprous chloride あるいは copper chloride.

3.2 実践トレーニング 構文150

84. □□ The transition-metal complex **catalyzed the reaction of A with B to give** product **C**.

その遷移金属錯体は，**A**と**B**との反応で触媒作用を示し，化合物**C**を与えた．

Point 化学反応を記述する基本構文の例．6.2.2 を参照．

85. □□ Oxidative addition **is one of the most important** elementary reactions.

酸化的付加は，最も重要な素反応のうちの一つである．

酸化的付加反応

A–B + M ⟶ A–M–B   （例） PhBr + Pd(0) ⟶ Ph–Pd–Br

- 金属の形式酸化数が正の方向に変化
- Pd(0)価錯体がPd(II)価錯体に変化
- 「A-BのMへの酸化的付加」あるいは，「MのA-B結合への挿入」ともいう．ただし，「MのA-Bへの酸化的付加」とはいわない
- 逆反応はA-BのA-M-Bからの還元的脱離

86. □□ Kinetic resolution is a reaction by which one of the enantiomers **is selectively converted into** a different compound.

速度論的分割とは，鏡像体のうちの一つが**選択的に**異なる化合物に**変換される**反応のことをいう．

速度論的分割

| R体 | S体 | + A →(cat. ML*) | R体 | + △B

出発はラセミ体   （S体 とAの反応で生成する）

不斉触媒存在下で，ラセミ体原料のS体（あるいはR体）だけを試薬Aと反応させることができると，キラルな未反応原料のR体（あるいはS体）とキラルな化合物Bが得られる．このような方法を速度論的分割という．

cat. = catalyst, M = metal, L* = 不斉配位子

87. ☐☐ The co-catalyst **is essential for** completion of the reaction. | 反応を完結させるためには，その共触媒が必須である．

Point completion は「終了」という抽象名詞として使われており，無冠詞で用いる．

88. ☐☐ This interaction **exerts** considerable **influence on** the conformation of the ligand. | この相互作用は，配位子のコンフォメーションにかなり影響をおよぼす．

Point ligand は［ligənd］と［laigənd］の2通りの読み方がある．

89. ☐☐ This unique feature **makes** this new catalyst **versatile**. | この独特の特徴が，この新触媒を**汎用性のあるもの**にしている．

Point SVOC（C は名詞，形容詞，過去分詞）の構文．「O を C にする」の意味．

90. ☐☐ Metal clusters are attractive **candidates as** catalysts **for** efficient C-H activations. | 金属クラスターは，触媒として，効率的な C-H 活性化の魅力的な**候補**である．

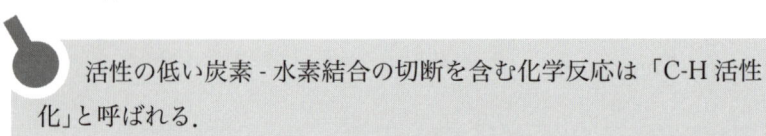

> 活性の低い炭素 - 水素結合の切断を含む化学反応は「C-H 活性化」と呼ばれる．

Point 「～に対する候補」というときは，candidate for ～とする．さらに，「～として～に対する候補」というように主語の範囲を限定したい場合は，candidate as ～ for ～とする．「～の候補」という日本語から考えると，candidate of ～ や candidate as ～ も可能に思えるが，英語では，単語によりその後に使える前置詞が限定されていることがある．英文を自作するときは，辞書の「英和活用」の用例を見て，どんな「名詞＋前置詞」の組合せが可能か調べる習慣をつけよう．そのほか effect of A on B（A の B に対する影響）にも注意．

Keywords

- be out of control　制御不能で
- conductivity　伝導度
- enhance　高める
- dramatically　劇的に
- conjugation　共役
- extend　拡張する
- significantly　著しく
- promote　促進する
- cupric chloride = copper dichloride　塩化第二銅（CuCl$_2$）
- cuprous chloride = copper chloride　塩化第一銅（CuCl）
- transition-metal　遷移金属
- complex　錯体
- catalyze　触媒作用を示す
- oxidative　酸化の
- elementary　基本の
- resolution　分割
- enantiomer　エナンチオマー，鏡像
- selectively　選択的に
- convert　変換する
- co-catalyst　共触媒
- essential　必須の
- completion　完了
- interaction　相互作用
- exert　働かせる
- considerable　かなりの
- influence　影響
- conformation　コンフォメーション，立体配座
- ligand　配位子
- unique　独特の
- feature　特徴
- catalyst　触媒
- versatile　用途の広い
- cluster　クラスター
- attractive　魅力的な
- candidate　候補
- efficient　効率のよい，有効な

#61 Lesson 10

91. ☐☐ Oxygen often **interferes with** the activity of the catalyst.

酸素は，しばしばその触媒の活性に悪影響をおよぼす．

Point この文での oxygen は物質名詞扱い．ただし，「酸素原子」という意味の oxygen は可算名詞扱いとする．

92. ☐☐ Two phosphines are equivalent in the ^{31}P NMR spectrum at room temperature **due to** the facile exchange reaction.

二つのホスフィンは室温で容易に交換反応するため，^{31}P NMR スペクトルでは等価である．

93. ☐☐ We **want to** propose a new process for the production of ethylene.

われわれは，エチレン生成の新しい工程について提案したい．

> **Point** want to ～ は自分の意見を主張するニュアンスがある場合があり，impolite な（無礼な）表現となりうるので注意．
> **Point** + to 不定詞をとる代表的な動詞：want, decide, promise.
> **Point** ethylene は具体的な一つの化合物名を指す物質名詞で不可算名詞．例文 1，32，35，48 を参照．

94. ☐☐ We **wish to** propose a novel process for the production of ethylene.

われわれは，エチレン生成の新しい工程について提案したい．（やや格式ばった表現）

95. ☐☐ We **would like to** propose a clean process for the production of ethylene.

われわれは，エチレン生成のクリーンな工程について提案させていただきたい．（より丁寧な表現）

96. ☐☐ Brass **consists of** copper and zinc.

真鍮は，銅と亜鉛から成る．

> **Point** 「～から成る」
> 　1. **A** consists of **B** and **C**. （受け身不可，例文 106 参照）
> 　2. **A** is comprised of **B** and **C**. （例文 104 参照）＝ **B** and **C** comprise **A**.
> 　3. **A** is constituted of **B** and **C**. ＝ **B** and **C** constitute **A**.
> 　4. **A** is composed of **B** and **C**. ＝ **B** and **C** compose **A**.
> 　5. **A** is made up of **B** and **C**.
> 　**A** ＝ **B** ＋ **C** の表現であるが，関係をまちがえないように，英文を書くときには，辞書を引きよく確かめる必要がある．

97. ☐☐ Recently, we have been **focusing on** electrically conductive (conducting) polymers.

最近，われわれは電気伝導性ポリマーに注目している．

98. ☐☐ Polyesters **as well as** nylons **are exploited for** the preparation of synthetic fibers.

ナイロンと同様，ポリエステルは合成繊維をつくるのに利用される．

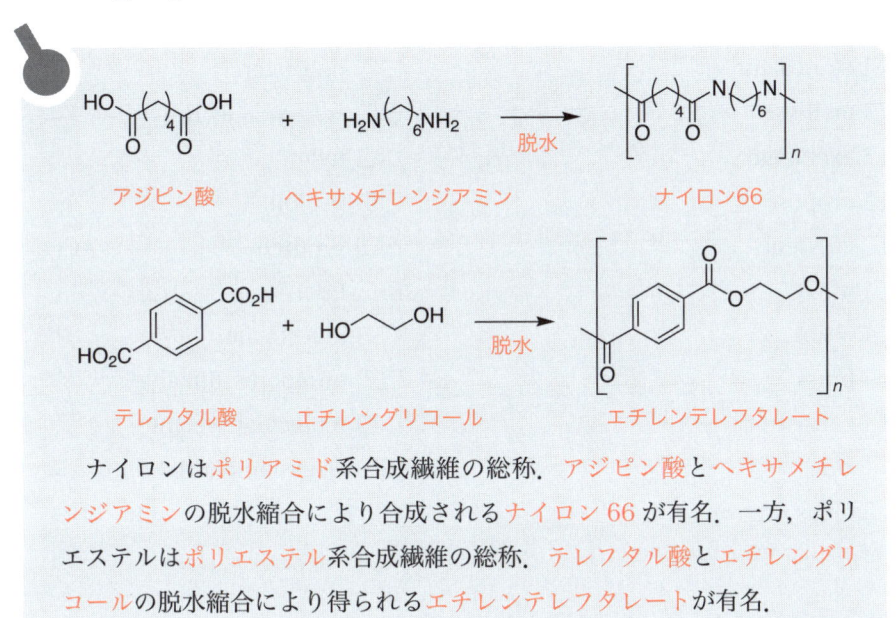

ナイロンはポリアミド系合成繊維の総称．アジピン酸とヘキサメチレンジアミンの脱水縮合により合成されるナイロン66が有名．一方，ポリエステルはポリエステル系合成繊維の総称．テレフタル酸とエチレングリコールの脱水縮合により得られるエチレンテレフタレートが有名．

Point この文章では，polyester と nylon は一つの化合物でなく，いろいろなポリマーを指す可算名詞として扱われており，複数形になっている．例文 1, 32, 35, 48, 93 を参照．

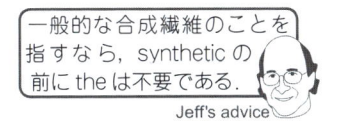

> 一般的な合成繊維のことを指すなら，synthetic の前に the は不要である．
> Jeff's advice

Point **A** as well as **B** が主語のときの動詞は **A** の人称・数と一致させるのが原則．ただし，**B** が複数のとき，あるいは as well as が and と同様の意味をもつときは複数形動詞が使われる．

99. ☐☐ We have **investigated** photocrosslinkable polymers.

われわれは，光架橋性ポリマーについて研究している．

100. The molecular weight of the polymer **does not follow** a Gaussian distribution.

そのポリマーの分子量は，ガウシアン分布に従わない．

Keywords

interfere with	妨げる	electrically	電気的に
activity	活性	conductive	伝導性の
phosphine	ホスフィン	polymer	ポリマー
equivalent	等価の	polyester	ポリエステル
facile	容易な	A as well as B	Bと同様にA
exchange	交換	nylon	ナイロン
propose	提案する	exploit	利用する
ethylene	エチレン	synthetic	合成の
novel	新規な	fiber	繊維
clean	クリーンな	investigate	研究する，調査する
brass	真鍮	photocrosslinkable	光架橋性の
consist of ～	～から成る	molecular weight	分子量
zinc	亜鉛	Gaussian distribution	ガウシアン分布
focus on ～	～を重点的に取り扱う		

#62 Lesson 11

101. **The purpose of this study is to clarify** the molecular weight distribution of this polymer.

この研究の目的は，このポリマーの分子量分布を解明することである．

Point clarify 以外の重要類義語：elucidate, reveal, disclose（「閉まっているのを開ける」から），demonstrate, unveil（「ベールを取る」から）．これらの単語はまとめて覚えておくとよい．

102. Here, we **report** the preparation of polypyrrole films.

ここでは，われわれはポリピロールフィルムの調製法について報告する．

3.2 実践トレーニング 構文150　67

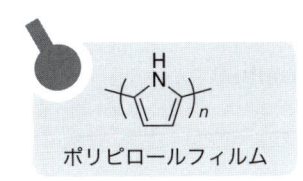

ポリピロールフィルム

Point　「～について報告する」は，report ～と report on ～がある．

> report ～のほうが直接的で，私ならこちらの表現を使う．
> Jeff's advice

103. □□ We synthesized several polymers **bearing** functional groups.

われわれは，官能基**をもつ**いくつかのポリマーを合成した．

104. □□ Woods **are comprised of** cellulose and lignin.

木は，セルロースとリグニン**から成る**．

> セルロースは，植物の細胞壁の主成分の多糖類で地球上最多の炭水化物．一方，リグニンは導管・繊維など木質化を起こす高分子化合物で，木材に20〜30%含まれる．地球温暖化・エネルギー問題の解決に向けて，これらを簡単に分解して容易に利用できるエネルギー源へと変換する方法が研究されている．

Point　wood が可算名詞扱いになっていることから，「いろいろな種類の木」という意味が含まれている．一方，cellulose と lignin はともにポリマーであるが，物質名詞扱いされており，それぞれ一つの化合物であると扱われている．例文 1, 32, 35, 48, 93, 98 参照．

105. □□ Coal **has been replaced by** petroleum.
    = Petroleum has replaced coal.

石炭は石油に**取って代わられている**．

Point　類似表現：Petroleum substituted for coal.

Chapter 3 化学に頻出する基本構文を覚えよう

106. ☐☐ The molecule in soap consists of **hydrophobic and hydrophilic moieties.**

セッケンに含まれる分子は，**疎水性と親水性の部分からなる．**

107. ☐☐ Such a soap **is called** an amphiphilic surfactant.

そのようなセッケンは，両親媒性界面活性剤**と呼ばれる．**

108. ☐☐ The efficiency of the drug in humans **has not been demonstrated** *in vivo*.

人体でのその薬の有効性は，**生体内では実証されていない．**

Point *in vivo* は「生体内の」の意. *in vitro* 「試験管内で」(p.118)とセットで覚えよう．

109. ☐☐ Histamine **is a chemical** found in normal tissues and blood.

ヒスタミンは，正常の組織や血液にもみられる化学物質である．

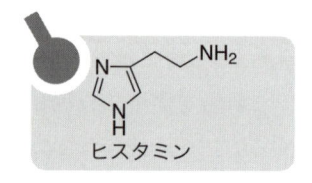

ヒスタミン

Point chemical は，「化学の」(形容詞)のほか，fine chemicals「ファインケミカルズ」のように名詞で「化学薬品」という意味もあることに注意．

Point histamine は具体的な化合物名であり，不可算名詞．例文 104 を参照．

110. ☐☐ The conductivity of a doped polyacetylene film **was comparable to** that of a Nichrome™ wire.

ドープしたポリアセチレンフィルムの伝導性は，ニクロム線のそれに**匹敵した．**

Point be comparable to ~で「~に匹敵する」の意味．

Point wire は一般には不可算名詞であるが，個々の線を指すときは可算名詞扱い．

Keywords

☐☐ purpose	目的	☐☐ drug	薬
☐☐ clarify	解明する	☐☐ demonstrate	実証する，証明する
☐☐ polypyrrole	ポリピロール	☐☐ in vivo	生体内で，インヴィヴォで
☐☐ film	フィルム	☐☐ histamine	ヒスタミン
☐☐ synthesize	合成する	☐☐ chemical	化学薬品，化学物質
☐☐ bear 〜	〜をもつ	☐☐ normal	正常の
☐☐ be comprised of 〜	〜から成る	☐☐ tissue	組織
☐☐ cellulose	セルロース	☐☐ blood	血液
☐☐ lignin	リグニン	☐☐ conductivity	伝導性
☐☐ replace	取って代わる	☐☐ dope	ドープする，不純物を添加する
☐☐ petroleum	石油		
☐☐ soap	セッケン	☐☐ polyacetylene	ポリアセチレン
☐☐ hydrophobic	疎水性の	☐☐ be comparable to 〜	〜に匹敵する
☐☐ hydrophilic	親水性の		
☐☐ moiety	一部分	☐☐ Nichrome™ wire	ニクロム線（ニクロム：NiとCrを主体とする合金）
☐☐ amphiphilic	両親媒性の		
☐☐ surfactant	界面活性剤		

#63 Lesson 12

111. ☐☐ The **high** price **is** a **huge drawback** to using this material.

値段の高いことが，この材料を使うことの大きな**欠点**である．

112. ☐☐ Substances that **cannot stand** high temperature **cannot be employed** for that purpose.

その目的には，高温に**耐えない**物質は**使用できない**．

Point 論文中では can't や wasn't のような短縮形は通常用いられない．

113. ☐☐ Optical disks (英：discs) can **be categorized into** the following three groups.

光学ディスクは，次の三つのグループに分類できる．

Point 「～に分類される」は，次の三つの表現を覚えよう！
be categorized into ～（アクセント位置注意）
be classified into ～（アクセント位置注意）
fall into ～

114. ☐☐ We have been **exploring** new materials with innovative functions.

われわれは，革新的な機能をもつ新素材を調査している．

Point explore for ～は，「～を探し求めて調査する」という意味になり，他動詞の explore とはニュアンスが異なる．

115. ☐☐ Similar methodology has already **been disclosed in** a patent.

同様の方法論が，すでにある特許のなかで開示されている．

Point methodology は，a set of methods and principles used to perform a particular activity の意．

> methodology は不可算名詞扱いし，冠詞をつけない．
> Jeff's advice

116. ☐☐ The study **provided conclusive proof** that DNA is a genetic material.

その研究は，DNA が遺伝物質であるという確証をもたらした．

117. ☐☐ Liquid crystals **were promising as** a material for display devices in those days.

当時，液晶は表示装置の材料として有望視されていた．

118. ☐☐ **According to the plan**, the synthesis of compound **A** will be completed next month.

その計画に従えば，化合物 **A** の合成は来月完了する．

119. ☐☐ Such an accident was **hardly** conceivable.

そのような事故は，ほとんど想像できなかった．

120. ☐☐ The explosion **is attributable to** the condensation of oxygen.

爆発したのは，酸素が液化した**ため**である．

> 外気にさらしたまま，液体窒素のなかにコールドトラップを入れておくと，空気中の酸素がトラップのなかに薄青色の液体となって凝縮する．このトラップを密閉系にして取りだすと爆発する危険性があるので注意．

Keywords

☐☐ price	価格
☐☐ huge	大きな，巨大な
☐☐ drawback	欠点
☐☐ material	物質
☐☐ substance	物質
☐☐ stand	耐える
☐☐ optical	光学の
☐☐ disk	ディスク
☐☐ be categorized into ～	～に分類できる
☐☐ following	次の，以下の
☐☐ explore	調査する
☐☐ innovative	革新的な
☐☐ function	機能
☐☐ methodology	方法論
☐☐ disclose	開示する，解明する
☐☐ patent	特許
☐☐ conclusive	決定的な，疑う余地のない
☐☐ proof	証拠
☐☐ genetic	遺伝(子)の
☐☐ liquid crystal	液晶
☐☐ promising	将来有望な
☐☐ display	ディスプレー，表示
☐☐ device	装置，デバイス
☐☐ in those days	そのころは
☐☐ according to ～	～によれば
☐☐ plan	計画
☐☐ be completed	完成する
☐☐ accident	事故

Chapter 3 化学に頻出する基本構文を覚えよう

□□ conceivable 想像できる
□□ explosion 爆発
□□ be attributable to ～ ～が原因である
□□ condensation 液化，凝縮

#64 Lesson 13

121. □□ These results **were unpredictable from** the working hypothesis.
これらの結果は，その作業仮説からは予期できなかった．

122. □□ That rule **cannot be applied to** this reaction.
あの規則は，この反応には適用できない．

123. □□ The fact **is worth paying particular attention to** because of its implications on the validity of the assumption.
その事実は仮定の妥当性について暗示しており，とくに注目に値する．

124. □□ Gasoline fractions of crude petroleum have boiling points **ranging from 40 to 200 ℃**.
未精製の石油のガソリン留分は，40 ℃ から 200 ℃の範囲の沸点をもつ．

125. □□ Einstein **played a prominent role in** the progress of modern science.
アインシュタインは，現代科学の進歩に顕著な役割を果たした．

Point play a role in ～ ～に役割を果たす
play a role of ～ ～の役割を果たす

126. □□ Kepler's law **was derived from** his careful examination of planets.
ケプラーの法則は，彼の惑星に対する注意深い考察から導きだされた．

Point Kepler's law は固有名詞扱いとなる．

したがって冠詞をつけない．
Jeff's advice

3.2 実践トレーニング 構文150

127. ☐☐ **With the advent of** the space age, a trip to the moon is no longer an unrealistic dream.

宇宙時代の**到来により**，月への旅行はもはや非現実的な夢ではなくなった．

128. ☐☐ His achievement **made a great contribution toward** (to) the chemical industry.

彼の業績は，化学産業に**偉大な貢献をした**．

129. ☐☐ We have **failed to** obtain good crystals of the protein.

そのタンパク質のよい結晶は得られていない．

130. ☐☐ The structure of the compound was **unambiguously determined** by X-ray crystallography.

その化合物の構造は，X線回折によりはっきりと決定した．

Point 和英辞典で「はっきりと」を引くと，clearly, sharply, vividly, distinctly, plainly, definitely などがヒットする．しかし，例文の文意のときは，unambiguously が最もよく使われる．一方，「仮に」は，tentatively を使うことが多い．

Point be determined by ＋ 方法

Point 一般的な方法を表す抽象名詞は通常，無冠詞で用いる．by the X-ray crystallography とはしない．

例文 54 と同じく，by X-ray diffractometer とはしない．
Jeff's advice

Keywords

☐☐ unpredictable	予測不可能な
☐☐ working hypothesis	作業仮説
☐☐ rule	規則，法則
☐☐ apply to 〜	〜に適用する
☐☐ attention	注目
☐☐ implication	暗示するもの，推測
☐☐ validity	有効性
☐☐ assumption	仮定
☐☐ gasoline	ガソリン
☐☐ fraction	留分
☐☐ crude	粗の，未精製の
☐☐ boiling point	沸点
☐☐ ranging from 40 to 200 ℃	40〜200℃の範囲の
☐☐ prominent	顕著な
☐☐ role	役割
☐☐ progress	進歩
☐☐ modern science	現代科学

Chapter 3 化学に頻出する基本構文を覚えよう

- ☐☐ Kepler's law　　ケプラーの法則
- ☐☐ be derived from ～
　　　　　　　　　　～から導きだされる
- ☐☐ examination　考察，調査
- ☐☐ planet　　　　惑星
- ☐☐ advent　　　　到来
- ☐☐ unrealistic　　非現実的な
- ☐☐ dream　　　　夢
- ☐☐ achievement　業績，達成
- ☐☐ contribution　貢献
- ☐☐ chemical industry　化学産業
- ☐☐ fail to ～　　　～に失敗する
- ☐☐ obtain　　　得る
- ☐☐ crystal　　　結晶
- ☐☐ protein　　　タンパク質
- ☐☐ structure　　構造
- ☐☐ unambiguously　明確に，はっきりと
- ☐☐ determine　　決定する
- ☐☐ X-ray　　　　X線
- ☐☐ crystallography　結晶学

#65 Lesson 14

131. ☐☐ The precise structure of the enzyme **remains ambiguous**.

その酵素の正確な構造は，いまだにはっきりしていない．

132. ☐☐ The product **crystallized as** orange needles.

その生成物は，オレンジ色の針状に結晶化した．

Point crystallized as ～ の as は，in the form of という意味を表す．

（Jeff's advice: この場合は，crystallized in ～よりも crystallized as ～のほうがよく使われる．）

133. ☐☐ These NMR spectra **were measured on** the spectrometer in the Instrumental Analysis Center.

これらのNMRスペクトルは，分析センターの分光計で測定されたものである．

134. ☐☐ You **should have taken** the ^{1}H NMR spectrum of the crude reaction mixture.

君は，クルード（未精製）の反応混合物の ^{1}H NMRスペクトルを測定しておくべきであった．

Point 実際には測定しなかったということを意味する．

135. ☐☐ The coupling constant *J* **is independent of** the applied magnetic field H_0.
　　結合定数 *J* は，かけられた磁場 H_0 **には依存しない**．

> 化合物の構造を解析するうえで役立つ NMR スペクトルの結合定数は，化合物固有の値である．

136. ☐☐ The present data **rigorously exclude** the mechanism they proposed.
　　このデータは彼らが提案したメカニズム**を厳密に除外している**．

Point data は datum の複数形で通常複数扱いである．しかし，データをひとまとめにとらえる場合には単数扱いも可．ただし a data とはしない．

137. ☐☐ This is one of the most practical reactions that **belong to** category **B**.
　　これは，**分類 B に属する**最も実用的な反応の一つである．

Point 先行詞が最上級の形容詞，the same, the only などで形容されているときには，必ず that を関係代名詞として用いる．

138. ☐☐ The temperature was maintained at –20 ℃ to **prevent degradation** of the chemicals.
　　その化学薬品の**品質低下を防ぐ**ために，温度を –20 ℃ に保った．

139. ☐☐ The researcher **collated** a huge amount of information from research institutes all over the world.
　　その研究者は世界中の研究機関からの莫大な量の情報**を比較対照した**．

140. ☐☐ **To (the best of) our knowledge**, nothing has been known about it.
　　われわれの知る限り，そのことについては何も知られていない．

Keywords

□□ precise	正確な	□□ magnetic field	磁場
□□ enzyme	酵素	□□ data	データ
□□ ambiguous	あいまいな，不明瞭な	□□ rigorously	厳密に
□□ crystallize	結晶化する	□□ exclude	除外する
□□ needle	針状結晶体	□□ practical	実用的な
□□ measure	測定する	□□ belong to ～	～に属する
□□ spectrometer	分光計，スペクトロメーター	□□ category	カテゴリー，部門
□□ instrumental	器機を用いる	□□ prevent	防ぐ
□□ analysis	分析	□□ degradation	分解，劣化
□□ center	センター，総合施設	□□ researcher	研究者
□□ coupling constant	結合定数	□□ collate	比較対照する
□□ be independent of ～	～に依存しない	□□ institute	研究所，(工科)大学
□□ applied	かけられた	□□ to (the best of) one's knowledge	～の知るかぎりでは

#66 Lesson 15

141. □□ You should **come up with** your own idea.　あなたは，あなた自身のアイデア**を提案すべきである**．

142. □□ The utility of this transformation **was exemplified by** those reactions.　この変換法の有用性は，それらの反応**によって例示された**．

143. □□ The fundamental concept of this methodology **is** actually **identical to** the conventional procedure.　この方法論の基本概念は，実際にはよくある手法**と同じである**．

Point be identical to ～で「～と同じである」の意味．

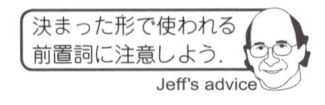

決まった形で使われる前置詞に注意しよう．
Jeff's advice

3.2 実践トレーニング 構文150

144. ☐☐ Their data **are** therefore **in conflict with** our conclusion.

それゆえ，彼らのデータはわれわれの結論**と矛盾している**．

145. ☐☐ These **aspects** must be considered.

これらの**局面**を考慮する必要がある．

146. ☐☐ The chemists **ascertained** what really happened.

その化学者たちは，何が実際に起こったのか**を確証した**．

147. ☐☐ This claim should **be rejected owing to** the lack of evidence.

この主張は，証拠不足**のため却下される**べきである．

148. ☐☐ The experimental data **were in good accordance with** the theoretical values.

この実験結果は理論値**とよく一致した**．

Point 通常 data は複数扱い（例文136参照）．

> 私は data を複数扱いする．
> Jeff's advice

149. ☐☐ This method has **been substantiated** by Prof. Terada.

その方法は，寺田教授により**実証された**．

150. ☐☐ We want to **convince you that** this transformation is very useful.

われわれは，**あなた方に**この化学変換がたいへん有用であること**を納得させたい**．

Keywords

☐☐ come up with	提案する
☐☐ idea	アイデア，理念
☐☐ utility	有用性
☐☐ exemplify	例示する
☐☐ fundamental	基本の，根本的な
☐☐ concept	概念
☐☐ actually	実際には
☐☐ be identical with ～	～と同じである
☐☐ conventional	通常の，月並みの
☐☐ therefore	それゆえ
☐☐ be in conflict with ～	～と矛盾している
☐☐ conclusion	結論
☐☐ aspect	側面，局面
☐☐ consider	考慮する
☐☐ ascertain	確かめる，確証する

☐☐ claim	主張	☐☐ theoretical	理論的な
☐☐ reject	却下する，不採用とする	☐☐ value	値
☐☐ owing to ～	～のために	☐☐ method	方法
☐☐ lack	不足	☐☐ substantiate	実証する，具体化する
☐☐ evidence	証拠	☐☐ convince	納得させる，確信させる
☐☐ be in accordance with ～	～と一致して		

Chapter 4 化学英語の文章に親しもう

いよいよ本書の最大の特色である，音声ファイルを最大限活用したリスニング，さらにはスピーキングによる化学英語の学習に本格的に取り組もう．ここでは化学の教科書として定評のある，オックスフォード大学出版局の "Chemistry Primers" シリーズを教材に取りあげる．さまざまな分野の基本的な英語にふれよう．

4.1 リスニング，スピーキングを用いた学習方法

① リスニング(テキストを見ないで)

まずはじめに音声ファイルを数回聞いて，どれぐらい聞き取れるかをチェックしよう．日本語に訳していては追いつかないので，英語をそのまま理解するように努めよう．

② 意味の確認

次に，テキストを見て文章と単語の意味をチェックしよう．意味を確認した単語は，チェックボックスに印をつけよう．

③ リスニング(テキストを見て)

続いて，テキストを見ながら何度か音声を聞いてみよう．

④ スピーキング

本文を音読してみる．Chapter 1 で学習した発音に注意しながら読もう．発音やアクセントの位置があいまいな単語は，それぞれ辞書で発音記号を調べて確かめよう．何度か音読したら，自分が読める限界のスピードで読んでみる．慣れないうちは詰まってしまうことが多いが，そこで諦めてはいけない．とにかく繰り返し読んでいると，次第に舌と口が動くようになってくる．

⑤ オーバーラッピング

④の時点でかなりくたくたになっているはずであるが，次に音声ファイルと自分の声を重ねて読んでみよう(巻頭「本書の特長と効果的な学習法」参照).

⑥ シャドーイング

オーバーラッピングができるようになったら，テキストを閉じてシャドーイングの練習をしよう．このとき，頭をリラックスさせて，意味のことなど考えず，とにかく耳に入ってくる音をオウム返しする気持ちで臨もう．

①と②は，あくまで英語学習のスタートにすぎない．シャドーイングができるようになるまで，根気よくじっくり時間をかけて繰り返し練習すること．相当なエネルギーと時間が必要となるが，この学習方法は「読み・書き・聴き・話す」すべての能力を伸ばす秘訣である．この Chapter のすべての英文を使ってこの学習方法を実践し，すべてシャドーイングできるレベルにまで学習すれば，英語力が総合的にアップしていることは確実であろう．もし，なかなかシャドーイングできるレベルにまで到達できないとしても，英語を声にだして学ぶ学習方法のよさが納得できたら，どんな英語の読み物でもよいから声にだして読むことを習慣づけよう．日に日に英語力が高まっていくはずだ．

各 Lesson 英文の OXFORD CHEMISTRY PRIMERS (OCP) シリーズ出典名
Lesson 1   "OCP 94: Foundations of Inorganic Chemistry" M.J.Winter & J.Andrew (2000)
Lesson 2   "OCP 9: Foundations of Organic Chemistry" G.M.Hornby & J.M.Peach (1993)
Lesson 3   "OCP 9: Foundations of Organic Chemistry" G.M.Hornby & J.M.Peach (1993)
Lesson 4   "OCP 81: Structure and Reactivity in Organic Chemistry" H.Maskill (1999)
Lesson 5   "OCP 45: Mechanisms of Organic Reactions" H.Maskill (1996)
Lesson 6   "OCP 63: Stereoselectivity in Organic Synthesis" G.Procter (1998)
Lesson 7   "OCP 9: Foundations of Organic Chemistry" G.M.Hornby & J.M.Peach (1993)
Lesson 8   "OCP 3: Organometallic Reagents in Synthesis" P.R.Jenkins (1992)
Lesson 9   "OCP 40: Foundations of Physical Chemistry" C.P.Lawrence, A.Rodger & R.G.Compton (1996)
Lesson 10 "OCP 85: Polymers" D.J.Walton & P.Lorimer (2000)
Lesson 11 "OCP 74: Supramolecular Chemistry" P.D.Beer, P.A.Gale & D.K.Smith (1999)
Lesson 12 "OCP 39: Photochemistry" C.E. & R.P.Wayne (1996)
Lesson 13 "OCP 98: Foundations of Chemical Biology" C.M.Dobson, A.J.Gerrard & J.A.Pratt (2001)
Lesson 14 "OCP 98: Foundations of Chemical Biology" C.M.Dobson, A.J.Gerrard & J.A.Pratt (2001)
Lesson 15 "OCP 13: Organometallic 2 - Complexes with transition metal-carbon π-bonds" M.Bochmann (1994)

4.2 実践トレーニング 教科書の英文 15

#67　Lesson 1　Atom　　　原子

Sometimes, it is useful to regard atoms as small billiard ball-like entities, but usually a chemist does need a more advanced view of atom structure. Atoms possess structure. They are not little, hard, featureless balls, although they are spherical. An atom consists of a positively charged nucleus surrounded by negatively charged electrons. Most of the volume of the atom is associated with the electrons. Although the radius of the nucleus is only about 0.01% of that of the atom, the total mass of the electrons is much less than that of the nucleus.

　　時には，原子を小さなビリヤードのボールのような存在と見なすことも役に立つが，通常化学者は，原子構造に関してより進んだ見方をすることが必要である．原子には構造がある．原子は球状ではあるのだが，小さくて固い，特徴のない球体というわけではない．原子は，負に帯電した電子に取り囲まれた，正に帯電した核から成っている．原子の体積のほとんどは，電子と関係している．核の半径は原子の半径のたった 0.01% ほどであるが，電子の総質量は，核の総質量よりずっと小さい．

Keywords

□□ atom	原子	□□ consist of 〜	〜から成る
□□ regard A as B	AをBと見なす	□□ positively	正に
□□ billiard	ビリヤード	□□ charged	帯電した
□□ 〜like	〜のような	□□ nucleus（複数形は nuclei）	核
□□ entity	存在物	□□ surround	取り巻く
□□ chemist	化学者	□□ negatively	負に
□□ advanced	高等な，進んだ	□□ electron	電子
□□ view	見解	□□ volume	体積
□□ structure	構造	□□ is associated with 〜	〜と関連がある
□□ possess	（性質などを）もつ		
□□ featureless	特色のない	□□ radius（複数形は radii）	半径
□□ spherical	球形の	□□ mass	質量

Point 13行目 than that of the nucleus は，than total mass of the nucleus と書き換えられる．

#68

Atoms are very small, with radii of the order of 100 pm, meaning that several million atoms could line up in a less than a millimeter long.

The nucleus contains one or more protons which from the chemist's point of view are positively charged indivisible particles. Apart from hydrogen, the nucleus also contains neutrons, also indivisible, which weigh about the same as protons, but are electrically neutral. The identity of the atom is defined by the number of protons within the nucleus. So the nucleus within all hydrogen atoms possesses only one proton, the nucleus within all helium atoms contains two protons, and so on. The nucleus is surrounded by a number of electrons.

原子はたいへん小さく，100 pm のオーダーの半径をもっている．つまり，数百万個の原子が 1 mm に満たない長さに一列に並ぶ小ささである．

核には一つ以上の陽子があるが，それは，化学者の視点からすると分割できない正に帯電した粒子である．水素原子を除いて，核は中性子をもっている．中性子はやはり分割できず，陽子とおよそ同じ質量をもつが，電気的には中性である．原子の同一性は，核の中にある陽子の数によって定義される．すなわち，すべての水素原子の中の核は，ただ一つだけの陽子を所有し，すべてのヘリウム原子の核は二つの陽子を含むといった具合である．核はいくつかの電子に取り囲まれている．

Keywords

☐☐ order	次数	☐☐ contain	含む
☐☐ line up	一列に並ぶ	☐☐ neutron	中性子
☐☐ millimeter	ミリメートル	☐☐ weigh	重さがある
☐☐ ～ long	～(数量を表す語)の長さの	☐☐ electrically	電気的に
		☐☐ neutral	中性の
☐☐ contain	含む	☐☐ identity	同一であること
☐☐ proton	陽子	☐☐ define	定義する
☐☐ indivisible	分割できない	☐☐ one and only ～	唯一無二の～
☐☐ particle	粒子	☐☐ helium	ヘリウム
☐☐ apart from ～	～は別にして	☐☐ and so on	～など
☐☐ hydrogen	水素		

Point 2行目の ~, meaning that ~は分詞構文．~, which means that ~と考えるとわかりやすい．

Point 5行目の ~ which from the chemist's point of view are ~（化学者の視点から考えると）は，化学では不可能だが，素粒子物理学（核反応）では分割できることを暗に示唆している．

Point 8行目のneutrons, also indivisible, は，neutrons (which are) also indivisibleと考えるとわかりやすい．

#69

Electrons are negatively charged but the magnitude of the charge is precisely the same as that upon a proton. The effect is that the atom is electrically neutral. While all atoms of one element possess the same number of protons, the nuclei of atoms of any one element may contain different numbers of neutrons. Atoms with the same number of protons but differing number of neutrons are called isotopes. So, all carbon atoms possess six protons and most contain six neutrons. The atomic mass of this form of carbon is 12 (6 + 6) and this isotope is denoted $^{12}_{6}C$.

電子は負に帯電しているが，電荷の大きさは陽子にある電荷と正確に等しい．その結果，原子は電気的に中性である．ある一つの元素のすべての原子は，まったく同じ数の陽子をもつが，いかなる一つの元素の原子の核も，異なる数の中性子を含む可能性がある．同じ数の陽子をもち，異なる数の中性子をもつ原子は同位体と呼ばれる．すなわち，すべての炭素原子は6個の陽子をもち，ほとんどが6個の中性子をもっている．この形式の炭素の原子質量は12 (6 + 6) であり，この同位体は $^{12}_{6}C$ と表示される．

Keywords

- ☐☐ magnitude　規模，大きさ
- ☐☐ precisely　正確に，精密に
- ☐☐ isotope　同位元素
- ☐☐ denote　示す

#70

A small number of carbon atoms (1.1 % in naturally occurring carbon) contain seven neutrons. These have an atomic mass of 13 and are denoted $^{13}_{6}C$. A very small proportion of carbon atoms, perhaps 1 in 10^{12} within living creatures, possess eight neutrons. Theses carbon atoms are $^{14}_{6}C$. The nuclei of carbon-14 atoms are not stable and decompose slowly, with a half-life of 5,715 years, to the nitrogen isotope $^{14}_{7}N$ in a process which releases a β-particle (an electron) from the nucleus and γ-radiation.

少数の炭素原子（天然に存在する炭素の1.1%）は7個の中性子を含む．これらは原子質量13であり，$^{13}_{6}C$と表記される．非常に少数の炭素原子は，だいたい生物体中に10^{12}個中1個の割合で，8個の中性子をもっている．これらの炭素原子は$^{14}_{6}C$と表記される．炭素14の核は安定ではなく，ゆっくりと分解するが，その半減期は5,715年で，核からβ線（電子）を放ち，γ線を放射する過程で窒素の同位体$^{14}_{7}N$になる．

Keywords
- naturally occurring　天然に存在する
- proportion　割合
- living creatures　生物体
- half-life（複数形は half-lives）　半減期
- decompose　分解する
- β-particle　β線
- γ-radiation　γ線

Point 10^{12} は，ten to the power of twelve と読む．ただし口語では，ten to the twelve や ten to twelve も使う．

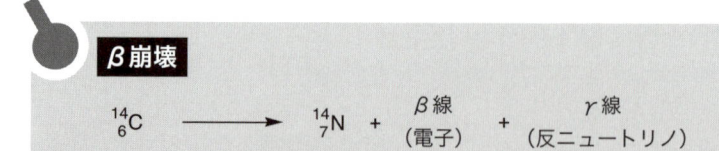

β崩壊

$$^{14}_{6}C \longrightarrow {}^{14}_{7}N + \underset{(電子)}{\beta 線} + \underset{(反ニュートリノ)}{\gamma 線}$$

放射性元素の原子核が電子と反ニュートリノをだし，中性子が一つ減少し，陽子が一つ増加する．$^{14}_{6}C$は大気上層で宇宙線をエネルギー源とした核反応により$^{14}_{6}CO_2$としてほぼ一定濃度で存在するが，生物が死ぬと$^{14}_{6}C$が生体内に取り込まれなくなる．このことを利用して年代測定に使用される．

#71 Lesson 2 Bond strength and lengths | 結合の強さと長さ

Two factors influencing the strength of a bond are its type and the sizes of the atoms involved. We define bond strength as the enthalpy change of the process $X\text{-}Y_{(g)} \to X_{(g)} + Y_{(g)}$. The term bond energy is often used for this. Table 1 shows that σ

結合の強さに影響を及ぼす二つの要因は，その結合の型（タイプ）と，結合に含まれる原子の大きさである．結合力はX-Y（気体）→ X（気体）+ Y（気体）の過程のエンタルピー変化であると定義される．結合エネルギーという言葉は，しばしばこのことに

Table 1

Bond	Length (nm)	Bond energy (kJ mol^{-1})	Type
C–C	0.154	346	σ
C=C	0.135	610	$\sigma + \pi$
C–O	0.143	358	σ
C=O (aldehyde)	0.122	736	$\sigma + \pi$

bonds are stronger than π bonds for C to C connections. The π bond between C atoms is therefore the easier part of the double bond to break and the chemistry of alkenes is dominated by this.

対して用いられる．表1は，炭素-炭素の結合において，σ結合がπ結合より強いことを示している．炭素原子間のπ結合は，それゆえ二重結合のなかでより切断されやすく，アルケンの化学はこのことによって支配される．

Keywords

- ☐☐ strength　力
- ☐☐ length　長さ
- ☐☐ factor　要因
- ☐☐ influence　影響を及ぼす
- ☐☐ type　型，種類
- ☐☐ size　大きさ
- ☐☐ involve　含む
- ☐☐ enthalpy　エンタルピー
- ☐☐ term　用語
- ☐☐ connection　連結，結合
- ☐☐ therefore　それゆえに
- ☐☐ chemistry　化学
- ☐☐ alkene　アルケン
- ☐☐ dominate　支配する

#72

For carbon to oxygen double bonds the reverse appears to be true. This could be one reason why there are more addition/elimination reactions in C=O chemistry than in C=C chemistry. Short bonds are usually strong ones.

炭素－酸素二重結合については逆のことが当てはまる．このことは，C=C の化学よりも，C=O の化学でより多くの付加／脱離反応が存在することの一因であろう．短い結合は，通常強い結合である．

Keywords

- oxygen　　酸素
- double bond　二重結合
- reverse　　反対
- appear to be ～　～と思われる
- true　　当てはまる
- addition　付加
- elimination　脱離
- reaction　反応

付加／脱離反応

$$R-C(=O)-OR' + R''O^- \xrightarrow{付加} \left[\begin{array}{c} R\ O^- \\ R'O\ \ OR'' \end{array} \right] \xrightarrow{脱離} R-C(=O)-OR'' + R'O^-$$

エステル交換反応のように，求核剤がカルボニル基に付加したのち，脱離基が脱離する置換反応のことを指す．

#73

The effect of atomic size on bond energy can be seen in the series C-halogen. The smaller the halogen atom, the closer the bonding nuclei can get to each other (Table 2).

All other things being equal, one would expect the C-I bond to be the easiest of the four to break. It is the comparative difficulty

元素の大きさが結合エネルギーに及ぼす効果は，一連の炭素－ハロゲン結合において見られる．ハロゲン原子が小さければ小さいほど，結合する核は互いにより接近できる（表2）．

ほかの状況がすべて同じであれば，C-I 結合が四つのなかで最も切断しやすいと予測できるであろう．C-F や C-Cl 結合切

of breaking C-F and C-Cl bonds that made chlorofluorocarbons (CFCs) a menace to the environment. Compounds such as CCl_2F_2 linger for years in the atmosphere and are unaffected by bacteria. There is evidence that they damage the ozone layer. They are now banned in the UK.

断が比較的困難であるため，クロロフルオロカーボン(CFCs)は環境に対する脅威となっている．CCl_2F_2 のような化合物は，大気中に何年もとどまり，また，バクテリアによる影響を受けない．それら（の化合物）がオゾン層を破壊する証拠がある．英国では，現在使用が禁じられている．

Table 2

Bond	Length (nm)	Bond energy (kJ mol^{-1})
C–F	0.138	452
C–Cl	0.177	339
C–Br	0.194	280
C–I	0.214	230

Keywords

□□ effect	効果	□□ compound	化合物
□□ halogen	ハロゲン	□□ linger	いつまでも残る
□□ expect	期待する	□□ atmosphere	大気(圏)
□□ comparative	比較的な	□□ be unaffected by ～	～に影響を受けない
□□ difficulty	困難		
□□ chlorofluorocarbon	クロロフルオロカーボン	□□ bacteria	バクテリア，細菌
		□□ evidence	証拠
□□ make A B	AをBにする	□□ damage	損壊する
□□ menace	脅威	□□ ozone layer	オゾン層
□□ environment	環境	□□ ban	禁止する

Point 2～4行目，The smaller the halogen (is), the closer ～は，「The 比較級, the 比較級」で「～するほどより～になる」の意味．

Point 5行目の All other things being equal は仮定法の分詞構文で，If all other things were equal, ～ と考える．

Point 7～10行目は，It is ～ that ～ の強調構文．

#74 Lesson 3 Hydrogen bonding 　水素結合

A special case of dipole/dipole attraction involves hydrogen bonded to a small highly electronegative atom (fluorine, oxygen, or nitrogen). This H atom is attracted to another N, O, or F atom; this second atom must carry a nonbonded electron pair. This is known as hydrogen bonding and is worth 10-40 kJmol^{-1}.

Hydrogen bonding is the strongest of the weak interactions and accounts for the high boiling points of alcohols compared with alkanes or ethers containing a similar number of electrons. Compare ethane (b.p. −89 ℃) H$_3$C-CH$_3$ with methanol (b.p. 64 ℃).

双極子−双極子引力の特別な場合は，小さくて電気陰性度の高い元素（フッ素，酸素あるいは窒素）に結合した水素原子を含む．この水素原子は，ほかのN，O，F原子——この2番目の原子は非結合性の電子対を必ずもつ必要がある——にも引きつけられる．これは水素結合と呼ばれ10〜40 kJ mol^{-1}程度である．

水素結合は弱い相互作用のなかでは最も強く，この水素結合により，よく似た数の電子を含むアルカンやエーテルと比較してアルコールの沸点が高いことが説明できる．エタン（沸点−89 ℃）とメタノール（沸点64 ℃）とを比較せよ．

Keywords

- □□ hydrogen bonding　水素結合
- □□ dipole　双極子
- □□ attraction　引力
- □□ electronegative　電気陰性の
- □□ fluorine　フッ素
- □□ nitrogen　窒素
- □□ nonbonded　非結合の
- □□ electron pair　電子対
- □□ interaction　相互作用
- □□ account for ～　～を説明する
- □□ boiling point　沸点
- □□ alcohol　アルコール
- □□ alkane　アルカン
- □□ ether　エーテル
- □□ ethane　エタン
- □□ methanol　メタノール

Point hydrogen bonding と hydrogen bond の違いについては Chapter 3 の例文 39 を参照．

10行目の weak interaction とは，静電相互作用，水素結合，π−πスタッキング相互作用など共有結合以外の相互作用を指す．

#75 Lesson 4 Catalysis / 触媒作用

Catalysis is the enhancement of the rate of a reaction by a compound (the catalyst) not generally present in the chemical equation which describes the reaction. Normally, a catalyst remains unchanged by the chemical reaction it catalyzes. It brings about the rate enhancement by providing a reaction pathway additional to the one which occurs in its absence. This additional pathway will have its own rate law, and the total rate of reaction in the presence of the catalyst is the sum of the catalyzed and uncatalyzed pathways as illustrated for a generic second-order reaction in Fig. 1.

触媒作用とは，その反応を記述する化学式には通常表れないある化合物（触媒）によって，反応の速度が増加することである．触媒は普通，触媒として作用するその化学反応では変化しないままである．触媒は，触媒なしで起こる反応経路に新しい経路を提供することによって反応速度増大をもたらすのである．この追加の経路は，それ自身の反応速度式をもち，触媒存在下での反応の全速度は，図1のように一般的な二次反応で示される，無触媒下の反応経路と触媒下の経路の和となる．

Keywords

- catalysis 触媒作用
- enhancement 増加すること
- rate 速度 (velocityは物体の速さ，speedは走る速さ)
- compound 化合物
- present ある
- chemical 化学の
- equation 式
- describe 記述する
- normally 普通は
- remain unchanged 変化しないままである
- bring about もたらす
- provide 提供する，供給する
- pathway 経路
- absence 存在しないこと
- law 法則，原理
- presence 存在
- sum 合計
- illustrate 図示する
- generic 一般的な
- Fig. 1 (= Figure 1) 図1
- second-order 二次の

Point catalysis は「触媒作用」という抽象名詞であり，無冠詞で用いる．
Point rate law は速度の法則ということで「速度式」となる．

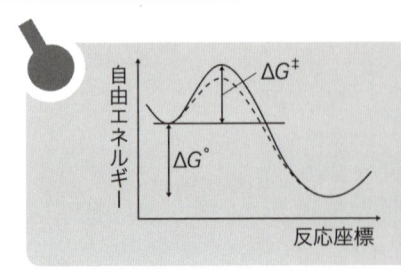

触媒は反応の活性化エネルギー $\Delta G^{\ddagger}$ を下げる．ただし触媒は，反応の ΔG° には変化を与えない（Chapter 3 例文 70 参照）．

#76

$$A + B \xrightarrow{k} C + D$$

rate of uncatalyzed reaction pathway
$= k[A][B]$

$$A + B \xrightarrow[X]{k_X} C + D$$

rate of reaction pathway catalyzed by X
$= k_X[X][A][B]$

Fig. 1 Uncatalyzed and catalyzed channels for a second-order reaction

For the reaction in Fig. 1,
total rate of reaction $= k[A][B] + k_X[X][A][B]$
$\qquad\qquad\qquad = (k + k_X[X])[A][B]$
Since [X] remains constant during the reaction, this rate law may be written
$\qquad$ total rate $= k_{obs}[A][B]$
where $\qquad k_{obs} = k + k_X[X] \qquad (1)$

無触媒の反応過程の速度は $k[A][B]$ であり，X によって触媒される反応過程の速度は，$k_X[X][A][B]$ である．

図 1 二次速度になる無触媒と触媒の経路

図 1 の反応に対しては，次のようになる．

$$全反応速度 = k[A][B] + k_X[X][A][B]$$
$$= (k + k_X[X])[A][B]$$

ここで，[X] は反応中一定であるので，この反応式は次のように書くことができる．

$$全速度 = k_{obs}[A][B]$$

ここで $\qquad k_{obs} = k + k_X[X] \qquad (1)$

Keywords

□□ channel　　経路
□□ constant　　定数，一定の

□□ k_{obs}　　（実測の）速度定数
　　　　　（obs は observation の略）

Point $(k + k_X[X])[A][B]$ は，open parenthesis k plus k_X times concentration of X close parenthesis times concentration of A and B と読む．

#77

The uncatalyzed reaction may be so slow that it is undetectable ($k \sim 0$), in which case the total reaction in the presence of the catalyst is effectively just the catalyzed reaction, $k_{obs} = k_X [X]$. This is very commonly the case for biological reactions involving enzymes as catalysts.

In Fig. 1, k_x is the catalytic constant for compound X and its effectiveness as a catalyst for the particular reaction is indicated by the ratio k_X/k. A catalyzed reaction is invariably catalyzed by a range of catalysts and each has its own catalytic constant; like all rate constant, each will depend upon the experimental conditions, e.g. temperature and solvent. The usual procedure for measuring a catalytic constant for a reaction represented by Fig. 1 is based upon eqn 1.

無触媒での反応は非常に遅く，検出できないかもしれない（$k \sim 0$）．その場合，触媒存在下での全体の反応は，実質的にただ触媒反応のみであり，$k_{obs} = k_X [X]$ となる．これは，酵素を触媒として含む生物学的反応では，ごく普通に見受けられる．

図 1 では，k_X が化合物 X に対する触媒定数で，その特定の反応に対する触媒としての有効性は k_X/k の比によって示される．触媒反応は，ある一連の触媒によって不変的に触媒され，それぞれはそれ自身の触媒定数をもっている．すべての速度定数のように，それぞれは，たとえば温度，溶媒といった実験条件に依存するであろう．図 1 で示した反応の触媒定数を測定する通常の方法は，式 1 に基づいている．

Keywords

☐☐ undetectable	検出されない	☐☐ a range of	一連の〜
☐☐ effectively	実際上	☐☐ depend upon	〜に依存する
☐☐ biological	生物学(的)な	☐☐ experimental	実験の
☐☐ enzyme	酵素	☐☐ condition	条件
☐☐ effectiveness	有効性	☐☐ solvent	溶媒
☐☐ particular	特定の，特別の	☐☐ procedure	手順，手続き
☐☐ indicate	示唆する	☐☐ measure	測定する
☐☐ ratio	比	☐☐ represent	表す
☐☐ invariably	一定不変に	☐☐ is based upon 〜	〜に基づく

#78

Values of k_{obs} are measured by the normal methods of kinetics for different (but constant) concentrations of X, then a graph of k_{obs} against [X] gives k_X as the gradient and the intercept at [X] = 0 gives k, the rate constant for the uncatalyzed reaction.

Note that in Fig 1, the catalyzed reaction is first order in catalyst, and k_X is a third order rate constant. If an example which is first order when uncatalyzed had been used, the catalyzed reaction would have been second order overall, but the method for determining k_X is exactly the same. The procedural details for measuring a catalytic constant are different if the catalyzed reaction is second order in the catalyst, but not difficult to work out.

k_{obs} の値は，濃度が異なった（ただし一定の）Xに対する通常の速度論の方法で測定される．そしてk_{obs}と[X]の関係を表すグラフは，傾きとしてk_Xを与え，[X] = 0での切片は，無触媒の反応の速度定数kを与える．

図1では，触媒反応では触媒に対して一次であり，k_Xは三次式の反応定数である．もし，無触媒の反応が一次である例を用いていたなら，触媒反応は全体で二次式となったであろう．しかし，k_Xを決定する方法はまったく同じである．もし，触媒反応が触媒に対して二次であると，触媒定数を求める手続きの詳細は異なるが，難しい問題とはならない．

Keywords

- ☐☐ value 値，価値
- ☐☐ kinetics 速度論，反応速度
- ☐☐ concentration 濃度
- ☐☐ graph グラフ，図
- ☐☐ against ～ ～に対して
- ☐☐ gradient （グラフの）傾き
- ☐☐ intercept （グラフの）切片
- ☐☐ overall 全部で
- ☐☐ detail 詳細（しばしば「詳細」という意味で複数形で使われる）
- ☐☐ work out うまくいく，計算する

#79 Lesson 5 Carbonyl compounds | カルボニル化合物

In the context of organic synthesis, reactions of carbonyl compounds are probably the most useful of all; they are also amongst the most varied. This is because carbonyl is both the functional group of simple aldehydes and ketones (whose properties may be modified by conjugated unsaturation), and also a principal component of a range of the more complex functional groups of carboxylic acid derivatives such as esters, amides, anhydrides, etc.

有機合成においては，カルボニル基の反応はおそらくすべてのなかで最も役に立ち，また最も変化に富むものの範疇に入るであろう．これは，カルボニルがシンプルなアルデヒドやケトン（これらの性質は共役不飽和によって変化をつけることができる）の官能基であり，エステル，アミド，酸無水物などのカルボン酸誘導体の広範でより複雑な官能基の主要構成要素であるからである．

Keywords

carbonyl compound	カルボニル化合物
context	背景，状況
organic	有機の
synthesis	合成
amongst	= among
functional group	官能基
aldehyde	アルデヒド
ketone	ケトン
property	特性
modify	変更する，修正する
conjugate	共役する
unsaturation	不飽和
principal	主要な
component	構成要素
carboxylic acid	カルボン酸
derivative	誘導体
ester	エステル
amide	アミド
anhydride	無水物
etc.	～など

Point etc. は etcetera の略記で，「～など」という意味を表す．

#80

The distinction between compounds in which the carbonyl group is not directly bonded to a potential leaving group, e.g. structure (a), and those in which it is, structures (b), will prove helpful. As we shall see, the former undergo reactions with nucleophiles which, overall, are usually either addition or condensation reactions; in contrast, compounds (b) generally react to give substitution of the group X.

Functional groups which include a carbonyl are very widespread in the molecules of nature. An understanding, therefore, of how the carbonyl group reacts is essential for the study of biological chemistry as well as organic synthesis.

たとえば (a) の構造のようにカルボニル基に直接，潜在的な脱離基が結合していない化合物と，構造 (b) のように直接結合している化合物とを区別するのは役立つことであろう．これから見ていくように，前者は求核剤と反応し，通常，全体として付加反応か縮合反応を起こす．対照的に，化合物(b)は通常，Xの置換反応を起こす．

カルボニルを含む官能基は，自然界の分子中のいたるところに存在する．それゆえ，どのようにカルボニル基が反応するかを理解することは，有機合成のみならず，生化学の研究にとっても必須である．

(a) X = H, alkyl, & aryl (aldehydes and ketones)
(b) X = Cl, OR', OH, NR'R'', SR', etc. (carboxylic acid derivatives)

Keywords

☐☐ distinction	区別	☐☐ addition	付加
☐☐ potential	潜在的な	☐☐ condensation	縮合
☐☐ leaving group	脱離基	☐☐ in contrast	対照的に
☐☐ prove (to be)〜	〜であるとわかる	☐☐ widespread	広く行きわたった
☐☐ undergo	（反応などを）受ける	☐☐ nature	自然，天然
☐☐ nucleophile	求核剤	☐☐ essential	必須の
☐☐ overall	すべてを考慮に入れて	☐☐ biological	生物学(上)の

Point A as well as B に続く動詞については，Chapter 3 の例文 98 を参照．

#81 Lesson 6 Si face and re face

For most of the time, we will be concerned with reactions which involve the formation of tetrahedral, or sp^3, carbon atoms within a molecular framework. Such reactions are widespread. The addition of a Grignard reagent to an aldehyde **1** is just such a reaction. When the nucleophilic group is different to the substituent attached to the carbonyl carbon, then two products **2** and **3** are formed which are enantiomers. The products are chiral.

si 面と re 面

（この章では）おもに分子骨格のなかに四面体あるいは sp^3 炭素原子の生成を含む反応について考えてみよう．そのような反応は，至るところにある．グリニャール試薬のアルデヒド **1** への付加反応は，まさにそんな反応の一つである．求核的な基が，カルボニル炭素に結合した置換基と異なるときは，エナンチオマーである二つの化合物 **2** と **3** が形成される．その生成物はキラルである．

Keywords

- si face and re face　si 面と re 面
- be concerned with 〜　〜に関心をもつ
- formation　形成
- tetrahedral　四面体の
- carbon　炭素
- atom　原子
- molecular　分子の
- framework　骨格
- widespread　広く行きわたった
- addition　付加
- Grignard reagent　グリニャール試薬
- aldehyde　アルデヒド
- nucleophilic　求核性の
- substituent　置換基
- attached　結合している
- carbonyl　カルボニル
- product　生成物
- enantiomer　エナンチオマー，鏡像異性体
- chiral　キラルの

#82

The carbon atom which is 'responsible' for the chirality is often referred to as a chiral center or stereocenter. Enantiomers have opposite absolute configurations, and these are given the descriptors *R* and *S*. The assignment of a stereocenter as either *R* or *S* follows from the Cahn-Ingold-Prelog (CIP) convention. Details of this convention can be found in any major undergraduate text on organic chemistry.

キラリティーの源となっている炭素原子は，しばしばキラル中心あるいはステレオ中心と呼ばれる．エナンチオマーどうしは反対の絶対配置をもち，これらは記号 *R* と *S* で表される．*R* あるいは *S* としてのステレオ中心の帰属は，Cahn-Ingold-Prelog(CIP)の取り決めに従う．この取り決めの詳細は，有機化学の学部生向けの主要なテキストに載っている．

Keywords

- be responsible for 〜 〜の原因である
- chirality キラリティー
- refer to A as B AをBと呼ぶ
- enantiomer 鏡像体
- opposite 反対の
- absolute configuration 絶対配置
- give A B AにBを与える （ここでは受動形になっている）
- descriptor 記述語
- assignment （化学構造などの）帰属
- stereocenter ステレオ中心
- follow 従う
- convention 協定 （会議，慣習という意味でも重要）
- undergraduate 学部学生の
- organic chemistry 有機化学

Point 10行目 on organic chemistry の on は「〜に関する」の意味．

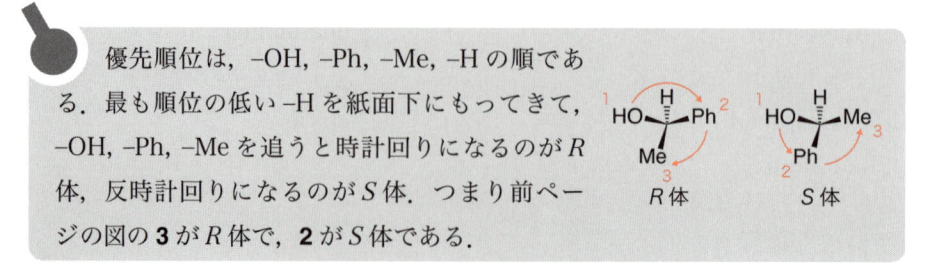

優先順位は，−OH, −Ph, −Me, −H の順である．最も順位の低い −H を紙面下にもってきて，−OH, −Ph, −Me を追うと時計回りになるのが *R* 体，反時計回りになるのが *S* 体．つまり前ページの図の **3** が *R* 体で，**2** が *S* 体である．

#83

In the nucleophilic addition to **1** which produces **2** and **3**, the enantiomers are formed in equal quantities, giving a racemic mixture. Enantiomer **2** would arise by addition from above the plane of the drawing, and **3** by addition from below. Stated another way, the two enantiomers arise by addition from the two faces of the carbonyl group. In order to avoid confusion as to which face is which ('upper' and 'lower' face will not do, as this depends on how the structure is drawn). The faces can be assigned a descriptor using a procedure similar to that which is used to assign absolute configuration (R and S).

化合物 **2** と **3** を生成する **1** への求核的付加では，エナンチオマーが同じ分量生成し，ラセミ混合物を与える．エナンチオマー **2** は，図の上面からの付加により生じ，**3** は下面からの付加によって生じる．別のいい方をすると，二つのエナンチオマーは，カルボニル基の二つの面からの付加により生じる．どちらがどちらであるか（上とか下というのは役に立たない．なぜならこれは構造をどのように書くかによるからである）という混乱を避けるため，面は，(R と S) の絶対配座を帰属する際に使用されるのと似た手続きで記号が割り当てられる．

Keywords

☐☐ quantity	量	☐☐ confusion	混乱
☐☐ racemic	ラセミの	☐☐ upper	上にある
☐☐ mixture	混合物	☐☐ lower	下の
☐☐ arise	生じる	☐☐ do	役立つ
☐☐ drawing	図面	☐☐ assign	割り当てる，帰属する

#84

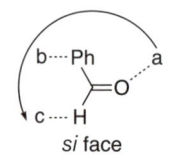

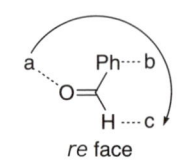

si face   *re* face

The carbonyl group is drawn in the plane of the paper, and the three groups attached are assigned a priority a, b, and c, using the same rules as for the assignment of absolute configuration. If the sequence a, b, c is anticlockwise it is the si face, and if it is clockwise it is the re face.

When the nucleophile attacks from the si face, enantiomer **2** is formed, whereas enantiomer **3** is formed when reaction takes place from the re face. These reactions take place at identical rates, because the transition states are equal in energy by virtue of them being enantiomers.

カルボニル基が紙面上に書かれている．そして結合した基に絶対配位の帰属に関するものと同じルールを使いa，b，cの優先権を割り当てる．もし，a，b，cの並びが反時計周りならsi面で時計周りならre面である．

求核剤がsi面から求核攻撃すると，エナンチオマー**2**が生成する．一方，re面から反応が進行するとエナンチオマー**3**が生じる．これらの反応は，同一の速度で進行する．なぜなら，それらはエナンチオマーどうしなので，遷移状態がエネルギー的に等しいからである．

Keywords

- □□ priority 優先順序
- □□ rule 規則
- □□ sequence 順序
- □□ anticlockwise (英)反時計周りの (米) counter clockwise
- □□ clockwise 時計周りの
- □□ identical まったく同じの
- □□ transition state 遷移状態
- □□ by virtue of 〜 〜のために
- □□ mirror plane 鏡面

#85 Lesson 7 Nucleophilic substitution / 求核置換反応

All the reactions in this and the following five sections follow similar mechanisms. If we write Nu: or Nu:⁻ for the nucleophile and X for the leaving group, the general mechanisms are:

この項と，次の五つの項のすべての反応は類似のメカニズムに従う．求核剤に対しNu: やNu:⁻ と書き，脱離基に対しXと書いたとき，一般的なメカニズムは次のようになる．

$$Nu: \curvearrowright C-X \longrightarrow Nu^+ - C + :X^-$$

or

$$Nu:^- \curvearrowright C-X \longrightarrow Nu - C + :X^-$$

The nucleophile's nonbonded pair forms the new Nu-C bond at the same time as the leaving group (X) goes off with the C-X bonding pair. We can now apply this general mechanism to many similar reactions.

求核剤の非結合性電子対が新しいNu-C結合をつくり，同時に脱離基XがC-Xの結合性電子対をもって離れていく．この一般のメカニズムを多くの類似の反応に適用することができる．

Keywords

☐☐ nucleophilic	求核性の	☐☐ mechanism	メカニズム
☐☐ substitution	置換	☐☐ nonbonded	非結合性の
☐☐ following	次に続く	☐☐ go off	離れる
☐☐ nucleophile	求核剤	☐☐ bonding	結合性の
☐☐ leaving group	脱離基	☐☐ apply A to B	AをBに適用する

nonbonding orbital（非結合性軌道）

MO（molecular orbital）は大きく次の三つに分類される．

- bonding orbital （結合性軌道）：この軌道に電子が収容されると結合次数が上がる．
- antibonding orbital （反結合性軌道）：この軌道に電子が収容されると結合次数が下がる．
- nonbonding orbital （非結合性軌道）：電子の授受によって結合次数は大きく変化しない．

#86 Lesson 8 Schlenk equilibrium

The reaction of an alkyl halide with magnesium to form a Grignard reagent, e.g. RMgBr is a fundamental organic reaction (Equation 1).

$$\text{Et-I} + \text{Mg} \longrightarrow \text{EtMgI} \qquad (1)$$

Almost all simple alkyl halides undergo this reaction to give the Grignard reagents which are stable at refluxing ether temperatures, but they do react with oxygen or moisture. Grignard reagents can react as bases or nucleophiles; however, in general they are less basic than the corresponding organolithium reagent. An important factor which governs the reactivity of Grignard reagents is their equilibrium with dialkylmagnesium and magnesium dibromide known as the Schlenk equilibrium (Equation 2).

$$2\text{EtMgBr} \rightleftharpoons \text{Me}_2\text{Mg} + \text{MgBr}_2 \qquad (2)$$
Schlenk equilibrium

シュレンク平衡

アルキルハライドとマグネシウムとの反応で，グリニャール試薬，すなわち RMgBr を生成することは，基本的な有機反応である(式1)．

ほとんどすべてのシンプルな(構造の)アルキルハライドはこの反応を起こし，エーテル還流温度下でも安定なグリニャール試薬を与えるが，それらは酸素や湿気とは反応する．グリニャール試薬は，塩基としても求核剤としても作用するが，一般に，対応する有機リチウム試薬よりも塩基性が低い．グリニャール試薬の反応性を決める一つの重要な要素は，ジアルキルマグネシウムとマグネシウムブロマイドとの間の平衡で，シュレンク平衡として知られている(式2)．

Keywords

- ☐☐ Schlenk equilibrium　シュレンク平衡
- ☐☐ alkyl halide　アルキルハライド
- ☐☐ magnesium　マグネシウム
- ☐☐ form　生成する
- ☐☐ Grignard reagent　グリニャール試薬
- ☐☐ fundamental　基本の
- ☐☐ reflux　還流する
- ☐☐ moisture　(大気中の)水分
- ☐☐ base　塩基
- ☐☐ in general　一般に
- ☐☐ basic　塩基性の，基本の
- ☐☐ corresponding　対応する，類似の
- ☐☐ organolithium　有機リチウム
- ☐☐ govern　支配する
- ☐☐ equilibrium (複数形は equilibria)　平衡
- ☐☐ dialkylmagnesium　ジアルキルマグネシウム
- ☐☐ magnesium dibromide　マグネシウムジブロマイド

#87

The presence of the Lewis acid $MgBr_2$ in solutions of Grignard reagents has a profound effect on their reactivity. In many reactions it is the coordination of $MgBr_2$ to the electrophile, such as an epoxide or carbonyl group, which makes the relatively poor nucleophilic Grignard reagent undergo addition and substitution reactions in high yield. The presence of $MgBr_2$ is also thought to catalyze the formation of Grignard reagents from alkyl halides and magnesium.

グリニャール試薬溶液中の，ルイス酸である $MgBr_2$ の存在は，反応性に大きな影響を与える．多くの反応で，エポキシやカルボニル基のような求電子剤への $MgBr_2$ が配位することにより，比較的弱い求核性のグリニャール試薬が，高収率で付加したり置換反応したりするのを可能にするのである．$MgBr_2$ の存在はまた，アルキルハライドとマグネシウムからグリニャール試薬の生成を触媒していると考えられている．

Keywords

☐☐ Lewis acid	ルイス酸	☐☐ relatively	相対的に
☐☐ solution	溶液	☐☐ poor	貧弱な
☐☐ profound	重大な	☐☐ nucleophilic	求核性の
☐☐ effect on ～	～への効果	☐☐ undergo	(反応などを)受ける，経る
☐☐ reactivity	反応性		
☐☐ coordination	配位	☐☐ reagent	試薬
☐☐ electrophile	求電子剤	☐☐ yield	収率
☐☐ epoxide	エポキシド		

#88 Lesson 9 Kinetics 速度論

We have already considered the hydrolysis of 2-chloro-2-methylpropane under S_N1 conditions where the following mechanism operates

$$(CH_3)_3CCl \rightarrow (CH_3)_3C^+ + Cl^- \quad (1)$$

If the concentration of 2-chloro-2-methylpropane is recorded as a function of time then it can be shown that the hydrolysis obeys first order kinetics

$$R = -\frac{d[(CH_3)_3CCl]}{dt} = k[(CH_3)_3CCl] \quad (2)$$

われわれはすでに，次に示すメカニズムが作用する 2-クロロ-2-メチルプロパンの S_N1 条件下での加水分解について考えてきた．

$$(CH_3)_3CCl \rightarrow (CH_3)_3C^+ + Cl^- \quad (1)$$

2-クロロ-2-メチルプロパンの濃度が時間に対する関数として記録されると，加水分解は一次速度式に従うことが示される．

$$R = -\frac{d[(CH_3)_3CCl]}{dt} = k[(CH_3)_3CCl] \quad (2)$$

Keywords

- □□ kinetics　　　（反応）速度論
- □□ consider　　　考慮する
- □□ hydrolysis　　加水分解
- □□ S_N1　　　　　S_N1 反応
- □□ condition　　　条件
- □□ following　　　以下の
- □□ operate　　　　機能する
- □□ mechanism　　メカニズム
- □□ record　　　　記録する
- □□ function　　　関数
- □□ obey　　　　　従う
- □□ first order kinetics　一次の反応速度（式）

2-クロロ-2-メチルプロパンの命名法であるが，2-chloro の c が 2-methyl の m よりアルファベットの順序が早いのでこの順に並べる．

#89

We have seen that for any system to react a certain energy — the activation energy — is required. In this example the source of this energy cannot be through collisions between the reactant molecules. Otherwise the process would be second order, rather than first order as is observed experimentally. Instead the energy required to overcome the activation barrier is supplied by occasional, highly energetic collisions from impacting solvent molecules.

われわれは，いかなるシステムでも反応を起こすためには，あるエネルギー—活性化エネルギー—が必要であることを見てきた．この例では，反応を起こすエネルギーの源は反応体分子間の衝突によるものではない．さもなければ，過程は実験で観測されるような一次式よりもむしろ二次式となるであろう．代わりに，活性化エネルギーを乗り越えるのに必要とされるエネルギーは，溶媒分子がときどき高エネルギー衝突をすることによって供給される．

Keywords

- certain　　ある，いくらかの
- activation energy　活性化エネルギー
- require　　必要とする
- example　　例
- source　　供給源
- collision　　衝突
- reactant　　反応物
- otherwise　　さもなければ
- process　　工程，プロセス
- A rather than B　BというよりもむしろA
- observe　　観測する
- experimentally　実験的に
- instead　　その代わり
- overcome　　乗り越える
- activation　　活性化
- barrier　　障壁
- supply　　供給する
- occasional　　たまの
- energetic　　エネルギーの，強力な
- impact　　強い衝撃を与える
- solvent　　溶媒

#90 Lesson 10 Polymer

Polymers are formed by linking large numbers of small molecules together. The term polymer derives from the Greek 'poly' meaning 'many' and 'mer' meaning 'units'. Perhaps the simplest of all macromolecules is polyethylene which can be viewed as simply an extension of the covalently bound micromolecule ethane. Here n represents the number of (CH_2-CH_2) units making up the polymer chain.

ポリマー

ポリマーは，数多くの小さな分子を一緒に結合することにより形づくられる．ポリマーという言葉は，ギリシャ語で「多い」を意味する「ポリ」，「単位」を意味する「マー」に由来する．おそらく，すべての巨大分子のなかで最も簡単なのは，ポリエチレンであり，それは，共有結合で結合した小さな分子エタンを単純に拡張したものとみなせる．ここで n は，ポリマー鎖をつくっているエチレン単位（CH_2-CH_2）の数を示している．

Keywords

□□ polymer	ポリマー	□□ extension	拡張したもの
□□ link	結合する	□□ covalently	共有結合の
□□ term	用語	□□ bound	結合した
（term A B で A を B と称する）		□□ micromolecule	ミクロ分子
□□ derive from ～	～から由来する	□□ ethane	エタン
□□ Greek	ギリシャ語の	□□ represent	意味する，表す
□□ unit	構成部分	□□ ethylene	エチレン
□□ macromolecule	巨大分子	□□ make up	構成する
□□ polyethylene	ポリエチレン	□□ chain	鎖
□□ be viewed as ～	～と見なされる		

#91

The value of n is termed the degree of polymerization $\overline{X}_n$. Since $\overline{X}_n$ could easily be a number as high as 10,000, then for a linear polymer with a repeat unit of relative molecular mass 28 (e.g. ethylene), this

n の値は，重合度 $\overline{X}_n$ と呼ばれる．$\overline{X}_n$ は容易に 1 万程度になり，直鎖状ポリマーでは，相対分子量 28（すなわちエチレン）の繰り返し単位をもち，これにより相対分子量（RMM）28 万のポリマーを生じること

would give a polymer of relative molar mass (RMM) 280,000. In other words

Molar mass of the chain = Molar mass of the repeat unit times $\overline{X}_n$

Other terms encountered are 'oligomers' meaning 'several units' (i.e. a very small polymer), trimer (three units, $n = 3$), dimer (two units, $n = 2$), and monomer, which is the basic unit from which a polymer is constructed.

になる．いい換えると

鎖の分子量 = 繰り返し単位の分子量 × 重合度($\overline{X}_n$)

よく見受けられるほかの用語には，'いくつかの単位'（すなわち非常に小さなポリマー）を意味する「オリゴマー」，三量体（三つの単位，$n = 3$），二量体（二つの単位，$n = 2$），およびポリマーをつくる基本単位であるモノマーがある．

Keywords

□□ value	価値，数値	□□ A times B	AかけるB
□□ degree	程度	□□ encounter	でくわす
□□ polymerization	重合	□□ oligomer	オリゴマー
□□ linear	直鎖上の	□□ trimer	三量体
□□ repeat	繰り返し	□□ dimer	二量体
□□ relative	相対的な	□□ monomer	単量体
□□ ethylene	エチレン	□□ basic	基本の
□□ in other words	言い換えると	□□ construct	組み立てる
□□ chain	鎖		

Point 1行目の The value of n is termed the degree of polymerization $\overline{X}_n$ は，We term the value of n the degree of polymerization.「われわれは，n の値を重合度と呼ぶ」の受動態．

Point 4行目，polymer は可算名詞．

#92 Lesson 11 Supramolecular chemistry | 超分子化学

For many years, chemists have synthesized molecules and investigated their physical and chemical properties. The field of supramolecular chemistry, however, has been defined as 'chemistry beyond the molecule', and involves investigating new molecular systems in which the most important feature is that the components are held together reversibly by intermolecular forces, not by covalent bonds.	長い間，化学者は分子を合成し，それらの物理的，化学的性質を調べてきた．しかし，超分子化学の分野は'分子を超える化学'として定義され，新しい分子システムを調べることを含む．その分子システムのなかでの最も重要な特徴は，構成要素が可逆的に，共有結合ではない分子間力で結びつけられている点にある．

Keywords

- supramolecular　超分子の
- chemist　化学者
- synthesize　合成する
- investigate　研究する，調査する
- physical　物理的な
- property（しばしば properties）（ものに固有の）性質
- be defined as ～　～と定義される
- feature　特徴
- hold together　一緒に結合する
- reversibly　可逆的に
- intermolecular　分子間の
- force　力
- covalent bond　共有結合

#93

Chemists working in this area can be thought of as architects combining individual covalently bonded molecular building blocks, designed to be held together by intermolecular forces(supramolecular glue), in order to create functional architectures (Fig. 1). Supramolecular chemistry is a multidisciplinary field, and therefore	この分野で研究している化学者は，機能的な建築物を創るために，分子間力（超分子接着剤）で結合するようにデザインされた，共有結合している個々の分子ビルディングブロックどうしを結合させる建築家であると見なすことができる(図1)． 超分子化学は，学際的な分野であり，それゆえ一連の基礎原理の理解が必要となる．このイントロでは，超分子科学の一般的な

requires a grasp of a range of basic principles. This introduction describes a generalized approach to supramolecular science and provides an indication of the wide ranging interests of chemists working in this area. Biological systems often provide inspiration, organic and inorganic chemistry are required for the synthesis of the pre-designed supramolecular components, and physical chemistry is used to fully understand their properties.

手引きについて述べ，この分野で研究している化学者たちに広く関心がもたれている点について述べる．生物学的なシステムはしばしば（われわれに）インスピレーションを与えてくれ，また有機および無機化学は，あらかじめ設計した超分子の構成要素の合成のために必要であり，物理化学は，それら（超分子）の特性を十分に理解するのに用いられる．

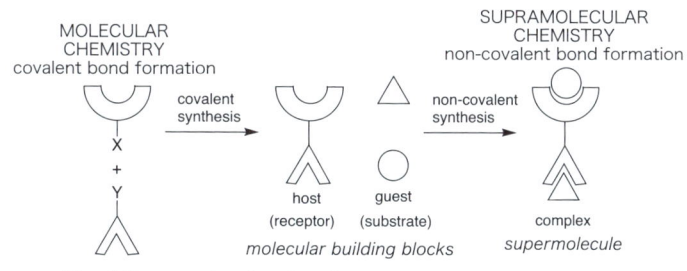

Fig. 1 Supramolecular chemistry

Keywords

- work　仕事をする，勉強する
- area　領域，分野
- architect　建築家
- combine　結合する
- individual　個々の
- building block　ビルディングブロック，（複雑なものを構成する）基礎単位
- design　設計する
- intermolecular　分子間の
- glue　接着剤
- create　創造する
- functional　機能的な
- architecture　建築

- multidisciplinary　学際的な
- therefore　それゆえ
- grasp　十分な理解
- a range of ～　一連の～
- principle　原理，法則
- introduction　入門書，序文
- generalized　一般化した
- approach　手引き，アプローチ
- provide　供給する
- indication　示唆するもの，兆候
- inspiration　着想の源，インスピレーション
- component　構成要素

#94 Lesson 12 Photosynthesis

Photosynthesis is perhaps the most important of the many interesting photochemical processes known in biology; not only was the evolution of the Earth's atmosphere dependent on it, but also animal life derives energy from the Sun, via photosynthesis, by eating plants. It is estimated that the total mass of organic material produced by green plants during the biological history of the Earth represents 1% of the Earth's mass, and that photosynthesis fixes annually the equivalent of ten times mankind's energy consumption.

From the point of view of organic synthesis, the overall process consists of the formation of carbohydrates by the reduction of carbon dioxide.

The essence of the process is the use of photochemical energy to split water and hence to reduce CO_2.

$$nCO_2 + nH_2O \xrightarrow{h\nu} (CH_2O)_n + nO_2$$

光合成

光合成は，おそらく生物学で知られている多くの興味深い光化学過程のなかで，最も重要であろう．というのは，地球を取り巻く大気の進化が光合成に依存しただけでなく，動物体は植物を食することにより光合成を通じて太陽からエネルギーを得ているのである．地球の生物学的歴史の期間において，緑色植物によって生産された有機物質の総質量は，地球の質量の1％に相当し，光合成は年間に人類のエネルギー消費の10倍に相当するエネルギーを固定化している．

有機合成の視点からすると，全体のプロセスは二酸化炭素の還元による炭水化物類の生成からなる．

そのプロセスの本質は，光化学のエネルギーを利用し，水を分解し，そしてCO_2を還元することである．

Keywords

☐☐ photosynthesis	光合成	☐☐ planet	惑星
☐☐ photochemical	光化学の	☐☐ fix	固定する
☐☐ biology	生物学	☐☐ annually	毎年
☐☐ evolution	進化	☐☐ equivalent	相当するもの
☐☐ atmosphere	大気	☐☐ mankind	人類
☐☐ life	生き物, 生物	☐☐ consumption	消費
☐☐ derive A from B	BからAを得る	☐☐ carbohydrate	（口語では carbo）炭水化物
☐☐ via 〜	〜を経由して		
☐☐ plant	植物	☐☐ reduction	還元
☐☐ estimate	見積もる	☐☐ carbon dioxide	二酸化炭素
☐☐ total	総計の	☐☐ essence	本質, 最も重要な特徴
☐☐ mass	質量	☐☐ split	分解する
☐☐ material	物質	☐☐ reduce	還元する
☐☐ represent 〜	〜に相当する		

Point 4行目は，not only が文頭に置かれたため倒置文になっている．

#95 Lesson 13 Cell –1–

Cells continually degrade organic compounds and synthesize new ones. The breakdown of complex compounds provides simple organic building blocks for the synthesis of other biological compounds as they are required. Chemical transformations of simple organic compounds are also used by cells to generate energy in a usable form. These chemical interconversions of 'metabolites' are the basis of metabolism. Of all the metabolites involved in these transformations, sugars and their simple derivatives play a central role.

This chapter describes the nature of chemical energy within cells and how this energy is harnessed from enzyme-catalyzed reactions. Cellular energy is related to the ability to facilitate dehydration chemistry in aqueous solution and the primary source of dehydrating power in cells, a phosphate derivative (ATP), is introduced.

細胞 1

細胞は，絶えず有機化合物を分解し，新しい有機化合物を合成する．複雑な化合物の分解は，生物学上必要とされるほかの化合物の合成のための簡単な有機ビルディングブロックを提供する．シンプルな有機化合物の化学的変換はまた，細胞によって，使用できる形態でエネルギーを発生するためにも活用される．これらの'メタボライト（代謝産物）'の化学的相互変換は，メタボリズム（新陳代謝）の基礎である．これらの物質変換に含まれるすべてのメタボライトのなかで，糖類とそれらのシンプルな誘導体が中心的な役割を果たしている．

この章では，細胞中の化学エネルギーの性質について，またこのエネルギーがどのように酵素触媒反応から利用されるかについて述べる．細胞のエネルギーは，水溶液中での脱水化学を促進する能力と関係しており，細胞中の脱水力の第一の根源であるリン酸エステル誘導体（ATP）が取りあげられている．

Keywords

☐☐ cell	細胞
☐☐ continually	断続的に（間に中断があってもよい）
☐☐ degrade	分解する
☐☐ breakdown	（消化による）分解
☐☐ complex (complex)	複雑な（「錯体」という意味もある）
☐☐ transformation	変換，トランスフォーメーション
☐☐ generate	発生させる
☐☐ usable	利用できる
☐☐ interconversion	相互変換
☐☐ metabolite	新陳代謝に必要な物質
☐☐ basis	基礎
☐☐ metabolism	代謝
☐☐ sugar	糖
☐☐ derivative	誘導体
☐☐ play a ~ role	~の役割を果たす
☐☐ nature	性質，本質
☐☐ harness	（自然の力などを）利用する
☐☐ enzyme	酵素
☐☐ cellular	細胞の
☐☐ be related to ~	~と関係がある
☐☐ ability	能力
☐☐ facilitate	促進する
☐☐ dehydration	脱水
☐☐ aqueous	水溶性の
☐☐ solution	溶液
☐☐ primary	基本的な，第一の
☐☐ phosphate	リン酸エステル

Point cell は「細胞」という意味のほかに，「電池」，「電解槽」，「(表計算の)セル」，「(刑務所の)独房」，「(ハチの巣の)巣室」などの意味がある．

Point continually に対して，(間に中断のない)「連続的な」は continuously．

Point 炭素の級数の「第一級の」も primary．

#96 **Lesson 14 Cell –2–**

Cell use many forms of energy, including mechanical energy (e.g. in muscle contraction) and electrical energy (e.g. in nerve signals), as well as chemical energy. However, all the energy-required processes of cells are related to chemical energy and this will provide the focus of our discussion. The nature of chemical energy within cells relates to the ability to turn processes that would normally be unfavorable into favorable ones; it is easiest to understand this concept by the use of examples.

Precisely constructed polymers are characteristics of life. The polymers common to all living systems correspond to the joining of monomer units with concomitant loss of water. They are known as condensation polymers, and include proteins derived from amino acids, polysaccharides derived from sugars and nucleic acids, the molecules associated with genetic information.

細胞2

細胞は，化学エネルギーと同様に，（たとえば筋肉の収縮で使われるような）機械的エネルギーや（神経シグナルで使われるような）電気的エネルギーを含む多くの形態のエネルギーを活用する．しかし，細胞でのエネルギーを必要とするすべての過程は化学エネルギーに関連しており，このことが議論の中心となる．細胞中の化学エネルギーの性質は，通常は不利な過程を有利な過程に変える能力と関連している．いくつかの例を使うことでこの概念を最も容易に理解できる．

精密に構築されたポリマーは，生命の特徴である．すべての生命システムに共通するポリマー（の生成）は，同時に水を失いながらモノマー単位が結合することに該当する．それらは，縮合ポリマーと呼ばれ，アミノ酸から誘導されるタンパク質，糖質から誘導される多糖類，遺伝情報に関連する分子である核酸などを含む．

Keywords

☐☐ form	形態	☐☐ characteristic	(しばしば〜s)特徴
☐☐ mechanical	機械的な	☐☐ common to 〜	〜に共通する
☐☐ muscle	筋肉	☐☐ correspond to 〜	〜に該当する
☐☐ contraction	収縮	☐☐ join	連結する
☐☐ electrical	電気に関する	☐☐ concomitant	同時に生じる
☐☐ nerve	神経	☐☐ loss	失うこと
☐☐ signal	信号	☐☐ condensation	縮合
☐☐ be related to 〜	〜と関係がある	☐☐ protein	タンパク質
☐☐ focus	(興味の)中心，焦点	☐☐ amino acid	アミノ酸
☐☐ discussion	議論	☐☐ polysaccharide	多糖
☐☐ relate to 〜	〜に関連する	☐☐ sugar	糖
☐☐ unfavorable	不利な	☐☐ nucleic acid	核酸
☐☐ concept	概念	☐☐ genetic	遺伝子の
☐☐ precisely	精密に		

Point 14〜15行目，The polymers common to all 〜は The formations of polymers that are common to all 〜と補って考えるとわかりやすい．

#97 Lesson 15 Bonding of alkenes to transition metals

Olefins bind to transition metals via their π orbitals, by donating electron density into an empty metal d-orbital. However, since olefins are weak bases, the bond has to be stabilized by another bonding contribution: the donation of electron density from the metal to the olefin, more precisely into its π* orbital which has the right symmetry for effective overlap with an occupied metal d-orbital.

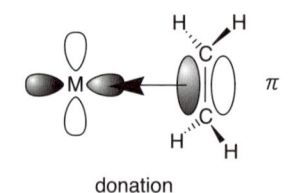

アルケンの遷移金属への結合

オレフィンはπ軌道を通じて，電子密度を金属の空のd軌道に供与することにより，遷移金属と結合する．しかし，オレフィンは弱い塩基なので，その結合は，ほかの結合の寄与により安定化させられる必要がある．それは，電子密度の，金属からオレフィンへの供与であり，より厳密にいえば，π*軌道への供与であり，電子を占有した金属のd軌道と効果的に重なるためのよい対称性がある．

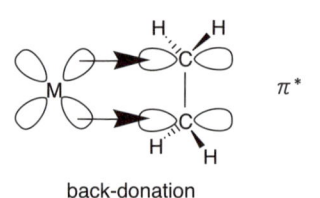

donation   back-donation

Keywords

bonding	結合
alkene	アルケン
transition metal	遷移金属
olefin	オレフィン
bind	結合する
via ~	~を経由して
orbital	軌道
donate (donate)	供与する
electron density	電子密度
empty	空の
d-orbital	d軌道
base	塩基
stabilize	安定化する（発音注意，aは二重母音[ei]）
contribution	寄与
donation	供与
precisely	正確に
right	好ましい
symmetry	対称性
effective	有効な
overlap	重なり
occupy	（電子を）占有する
back-donation	逆供与

#98

The metal therefore acts as both a Lewis acid (electron acceptor) and a Lewis base (electron donor) with respect to the olefin. This bonding concept was first proposed by Dewar and by Chatt and Duncanson ('Dewar-Chatt-Duncanson model') and has proved its value in understanding metal-alkene bonding over the years.

金属は，それゆえ，オレフィンに関してルイス酸(電子受容体)としてもルイス塩基(電子供与体)としても作用する．この結合の概念は，最初デュワー，チャットおよびデュンカンソン（デュワー-チャット-デュンカンソンモデルと呼ばれる）によって提案され，何年にもわたって金属−アルケン結合を理解するのに役立ってきた．

Keywords

- therefore　　それゆえ
- act as 〜　　〜としての機能を果たす
- Lewis acid　　ルイス酸
- electron　　電子
- acceptor　　受容体
- Lewis base　　ルイス塩基
- donor　　供与体
- with respect to 〜　　〜に対して
- concept　　概念
- propose　　提案する
- prove　　証明する
- value　　価値
- in understanding　　理解する際に (in + 〜ing)

Chapter 5 化学の論文や記事を読みこなそう

> Chapter 4 では代表的な教科書の例文を用いたリスニング，スピーキング，オーバーラッピング，シャドーイングの練習を通して，化学英語の文章に慣れ親しんだ．この Chapter 5 では，英文を速読する力を養うとともに，しっかりした日本語訳ができるように鍛錬をする．いろいろな形態の 7 つの英文を取りあげるが，まず日本語を介さず全文をざっと——できれば辞書を使わないで——読み，大意をつかもう．その後，辞書をフルに活用して，日本語らしい文章になるよう十分に注意して訳していこう．辞書なしで英文を速読する力と，あらゆる辞書を駆使してしっかりした日本語訳をつける力とが両方求められる．

5.1 英語長文の読解法

① 頭のなかを英語に切り替える

長い英語を速く読むときは，一文一文日本語に訳していてはいけない．英語の思考回路で理解する必要がある．できるだけ多くの英文を読む習慣をつけて，頭のなかを切り替えるトレーニングをしたい．

② 日本語らしい訳文にするには

一方，英語を日本語に訳すには，日本語の思考回路で考える必要がある．なぜなら，英語は「主語＋動詞（述語）」が基本であるが，日本語では述語は文章の最後に置かれるし，日本語ではふつう，能動形で表現するところを，英語では受身形で表現するほうが自然なことがあるなど，基本的な構文が異なるからである．研究室での雑誌（論文）の抄録会などで，英文から日本語の資料を作成するときには，「直訳でヘンテコな日本語」にならないようによく推敲したい．

5.2 実践トレーニング 英文和訳 7

Lesson 1　Kinetic Resolution

　The reactions using novel chiral acylation catalysts have been developed. The products have been easily separated by the kinetic resolution of racemic alcohol **A**, resulting in high enantioselectivities in most cases. The results indicated that H-bonding interactions between the catalyst and the substrate would be attributable to the high enantioselectivity in the present catalytic acylation. Additional study on the solvent dependence was also performed to clarify the reaction mechanism. The reaction can be applied to various alcohols with different substituents. Among examined, ethanolamine derivatives were found to be highly effective (up to S = 16.8) in particular.

- **Point** 1 行目の～ have been investigated は，「～が開発された」とするよりも「～を開発した」と能動態で訳すほうが日本語としては自然．
- **Point** 2 行目の kinetic resolution は，Chapter 3 の例文 86 を参照．
- **Point** 4 行目の be attributable to ～「～に起因する」は，Chaper 3 の例文 120 を参照．
- **Point** 5 行目，study on の on は，「～に関する」という意味．

Lesson 2　Catalytic Properties of RNA

　It has been known since the early 1980s that ribonucleic acids (RNAs) do not only participate in the flow of genetic information, but may also have catalytic properties. This discovery, which was honored with the Nobel Prize in 1989, was stimulated by the earlier hypotheses about the existence of a "RNA world" in which both the storage of genetic information and the control of chemical reactions were carried out by RNA. In recent years, the use of *in vitro* selection and evolution techniques has provide significant contributions to the exploration of the catalytic potential of RNA. While the exploration of the catalytic potential of RNA initially focused on reactions at the functional group present in RNA, such as acylation and alkylation, recent work has shown that some popular reactions, such as the Diels-Alder reaction, are possible with the use of dienes artificially linked to DNA.

　　　［平成 12 年度　大阪大学大学院工学研究科博士前期課程物質化学専攻 入試問題より］

- **Point** 1 ～ 2 行目には not only **A** but also **B**（**A** だけでなく **B**）の構文が使われている．
- **Point** 6 行目の *in vitro* については，Chapter 3 の例文 108 を参照．

Lesson 3 The Rare Earth Elements

In spite of their generic name, the rare earths are neither rare nor earths (metal oxides). They are metallic elements, and all but one are more rich in the earth than gold, silver, mercury or tungsten. The rare earth elements are actually ubiquitous and present in low concentration in virtually all minerals. But extracting the rare earths from common minerals would be costly. It is therefore fortunate that a handful of less common minerals do exist in which the rare-earth concentration is high enough to make the economical extraction of the elements possible, because the elements are key ingredients in the manufacture of numerous modern products. Cerium and erbium are components of high-performance metal alloys. Holmium and dysprosium are required for certain types of laser crystals. Neodymium is a key component of the strongest known permanent magnets, which have allowed new electric motor designs. Perhaps the most intriguing application of the rare earths is in the manufacture of ceramics that become superconducting magnet at significantly high temperatures.

Keywords

- □□ generic　　　属の
- □□ rare earth　　希土類元素
- □□ rich　　　豊富な
- □□ ubiquitous　　いたるところにある，遍在する
- □□ virtually　　事実上
- □□ costly　　（値段が）ひどく高い
- □□ do exist　（do は強調）実際に存在する
- □□ ingredient　　成分
- □□ cerium　　セリウム
- □□ erbium　　エルビウム
- □□ component　　成分
- □□ neodymium　　ネオジム
- □□ dysprosium　　ジスプロシウム
- □□ laser crystal　　レーザー結晶
  （レーザー光発生のために必要な結晶）
- □□ intriguing　　非常に興味をそそる
- □□ superconducting magnet　超伝導磁石

Point 11 行目の 〜, which 〜 は関係代名詞の非制限用法で，「そしてそれは」という意味．

Lesson 4　顕微鏡の使用説明書

Switch the microscope on and turn the light on. Next, place a slide on the microscope stage. Move the objective downwards until the sample is in focus. Adjust the sample at the center of the lens. Swing the high power lens into the position to view the sample at a higher magnification. When finished, remove the slide and turn the main switch off.

Keywords
- ☐☐ microscope　顕微鏡
- ☐☐ objective　対物レンズ
- ☐☐ magnification　倍率

Lesson 5　有機化学の教科書の序文

What enters your mind when you hear the words "organic chemistry"? Some of you may think, "the chemistry of life" or "the chemistry of carbon." Other responses might include "pre-med", "pressure", "difficult" or "memorization". Although formally it is the study of the compounds of carbon, the discipline of organic chemistry encompasses many skills that are common to other areas of study. Organic chemistry is as much a liberal art as a science, and mastery of the concepts and techniques of organic chemistry can lead to improved competence in other fields.

〔Susan McMurry, "Study Guide and Student Solutions Manual for McMurry's Organic Chemistry, 6th edition" Brooks/Cole, 2004 より〕

Point　5～6行目の as much **A** as **B** は，「**B** であるけれども **A** でもある」．ここは「一つの自然科学であるけれども教養科目でもある」の意．

Keywords
- ☐☐ pre-med　医学部へ進学するための課程
- ☐☐ memoration　暗記
- ☐☐ discipline　学問分野
- ☐☐ encompass　取り巻く
- ☐☐ liberal art(s)　一般教養科目
- ☐☐ mastery　精通
- ☐☐ competence　能力

Lesson 6　有機金属化学の教科書の序文

Much had happened in the six years since the first edition of this book appeared. The use of transition metals in organic synthesis has finally achieved general acceptance within the synthetic organic community, and it is rare to find complex total syntheses that don't involve at least a few transition-metal mediated key steps. In addition, industry has lost much of its aversion to using homogeneous catalysts. Finally, the uses of transition metals in the synthesis of polymers and materials and in solid-phase combinatorial chemistry have also increased dramatically.

[Louis S. Hegedus, "Transition Metals in the Synthesis of Complex Organic Molecules," University Science Books, 1999 より]

Keywords

□□ aversion　　嫌悪の情

Lesson 7　雑誌記事

FROM THE ACS MEETING
C-C BOND FORMATION IN ORGANIC CRYSTALS
Solid-state photochemical reaction offers alternative route to vicinal quaternary centers

Among the most challenging structural motifs in synthetic organic chemistry are adjacent quaternary centers. Miguel A. Garcia-Garibay, a chemistry professor at the University of California, Los Angeles, believes that carbon-carbon bond formation through photochemical decarbonylation of a crystalline ketone is a viable alternative way to create this motif, especially for compound of high value. Last month, Garcia-Garibay described applications of the reaction to total synthesis of complex molecules at the ACS national meeting in New York City.

Natural products with vicinal quaternary chiral centers are very difficult to prepare by solution chemistry, Garcia-Garibay said. On the other hand, if an appropriate substrate for photochemical decarbonylation can be prepared, the solid-state reaction is straightforward, forming the desired product in yields of up to 95% and enantiomeric excesses approaching 100%.

The idea is simple: Take a crystalline chiral acetone derivative with six different substituents and photochemically break the two bonds of the carbonyl group. When CO is removed, the resulting radicals then join to form a bond that maintains the absolute configuration of the substrate. "In general, radicals are not chiral," Garcia-Garibay explained. "But being in the chiral environment of the crystal, they will be configurationally trapped. And because the two electrons are very close to each other, they quickly form the desired bond."

[*Chem. Eng. News*, October 13, 2003, **81** (41), p.72 より.
Copyright 2003 American Chemical Society]

Keywords

- □□ alternative　既存のものに代わる
- □□ vicinal　隣接する
- □□ quaternary　第 4 級の
- □□ challenging　難しい，意欲をかきたてる
- □□ motif　モチーフ (図柄を構成する単位)
- □□ viable　見込みのある
- □□ ACS national meeting　ACS ミーティング (アメリカ化学会が主催する学会)
- □□ be trapped　トラップされる

Chapter 6 化学を英文で書き表そう

化学に関する英文を書く必要が生じるのはどんなときだろう？ 読者が大学生であれば，まず大学院入学試験で和文英訳問題を解く場面が思いつく．この場合は，自分の知識の範囲内で解答する必要がある．それ以外の状況ではどうか？ 学術論文であれ，学会発表の予稿原稿を書く場合であれ，辞書の力を借りて英文を書く機会のほうが，自力だけで書く機会よりも圧倒的に多いことがわかる．つまり，現実には，いかによい辞書を手元に置き，それらをうまく引きこなして伝えたいことを英語で書くかということが求められる．そのことを念頭に置いてこの Chapter では，まず英作文の一般的な注意点についてふれる．続いて，化学反応の英語表現についてまとめて扱う．さらに，100 の和文英訳問題を実際に解いてみる．

6.1 英作文を書くときの心得

① 動詞は自動詞か他動詞かを必ず確かめる

日本語では，「その問題を議論する」でも「その問題について議論する」でも間違いではない．しかし英語の場合は，動詞の使い方に注意する必要がある．「議論する」に動詞 discuss を使う場合，discuss は他動詞であり目的語をとる．つまり，discuss the problem で「その問題を議論する」の意味になる．しかし，discuss about the problem とすると文法的に誤りとなってしまう（一方，名詞の discussion は，discussion about 〜で，「〜に関する議論」となる）．使い慣れていない動詞を使って英文を書くときは，それが自動詞であるか他動詞であるかを必ず確認しよう．

② 名詞の冠詞について吟味する

冠詞，とくに定冠詞 the をつけるか否かを十分に考慮する必要がある．

▶ 定冠詞 (definite article) the は可算名詞にも不可算名詞にもつけることができる
 (1) 特定の物事を指している場合
 (2) 書き手だけでなく，聞き手もわかっている場合
▶ 化合物が物質名詞 (不可算名詞) であるか可算名詞であるかをチェックする
 Chapter 3 の例文 1, 32, 35, 48, 93, 98, 104, 109 を参照．
▶ 始めの文字の発音が母音ならば an となる
 an LED, an X-ray であり，a UV absorption spectrum (半母音 [ju:] で始まる) である．Chapter 3 の例文 41 を参照．
▶ same の前には必ず "the" を使う
 定冠詞 the は省略されることがあるが，same のときは省略しないのが慣例になっている．名詞を特定する形容詞がつくときも the をつけることが多い．Chapter 3 の例文 9 を参照．

【例】その反応はまったく同じ反応条件下で進行した．

→ The reaction took place under the same reaction conditions.

③ **主語が三人称単数であるかないかを常に意識する**
 つまり，三単現の s をつけるか，has と have のどちらを用いるべきかを常にチェックする．英語の基本であるが，実際にはこの点に関する誤りは非常に多い．

④ **できるだけシンプルな構文にする**
 to 不定詞や分詞構文を活用し，まとめたほうがよい文章は単文にする．

【例】分詞構文を使った文に書き換えると，

The X-ray crystallographic analysis was measured to determine the exact structure of the complex. The result showed that the compound had *cis*-configuration.

→ The X-ray crystallographic analysis was measured to determine the exact structure of the complex, showing that the compound had *cis*-configuration.

前文の結果を受けて，「その結果は〜であることを示している (show)，確証している (demonstrate)」と述べるとき，あるいは前文の内容の結果，「〜が得られた」と述べる場合など，「…, showing that 〜」「…, demonstrating that 〜」「…to give 〜」と分詞構文や to 不定詞で表現したほうが文章全体が締まってくる．

⑤ 能動態と受動態の表現のどちらが適当かを考慮する

日本語では，省略可能な場合，しばしば主語を省略するのに対して，英語では分詞構文などの特殊な場合を除き，主語を省略しないのが記述文の基本である．英語は動詞の特性上，日本語に比べて受動態が多用される（たとえば，I was surprised to 〜）．また，新聞記事や自然科学の論文では，人称代名詞を主語に用いた英語は口語調に受け取られるため，より客観性を高めるために受動態が多用される．とくに化学論文の実験項では，通常，人称代名詞を用いない表現を使う．

【例】Chapter 3，例文 8 の The reaction mixture was stirred at –78 ℃ for 30 min 〜. は，日本語では「(われわれは) 反応混合物を –78 ℃ で 30 分間撹拌した」のほうが自然であり，「われわれは」の部分はこの場合，表記では省略する．

それに対し，英語では主語を省略しないので，能動態にすると We stirred the reaction mixture at –78 ℃ for 30 min 〜. と人称代名詞が主語になってしまう．受動態にすることでこれが回避できる．

⑥ 英語の基本構文に忠実に

日本語の語順をそのまま英文にすると，とくに強調する必要のないときでも副詞句を文頭に置いた文章が連続してしまい，英語としては不自然になってしまう．「S + V + その他の要素」を基本構文にした文章をまず考えよう．

【例】「この反応によって化合物 A が得られた」をそのままの語順で英語にすると，

    By this reaction compound A was produced.

となり，by this reaction を強調した英文となる．まずは S (主語) + V (動詞) + 他の文の構成要素という英語の基本構文で英文を練ってみよう．

    → Compound A was produced by this reaction.

⑦ **同じ単語・構文を続けて近い文章のなかで使わない**

決して誤りではないが，ボキャブラリーの少なさを露呈するものであり，センスが悪いと見なされる．よくある例として，when を使った構文や，「生じる」という意味での give や afford の連続使用などがあげられる．

⑧ **適切な表現かどうかを吟味する**

たとえば，harness, utilize, use はいずれも「利用する」の意味がある．しかし，harness は，harness solar energy や harness the river's energy など，おもに（自然の力を）「利用する」の意味で使われ，ほかの単語とは用法が異なる．例文をよく見て，どんなニュアンスのときにその単語が使われているか理解する．

⑨ **文章では短縮形を使わない**

Chapter 3 の例文 112 や 122 ですでに学んだように，化学論文では wasn't, can't, they've など短縮形の使用は避け，was not, cannot（not only **A** but also **B** などのように not を強調したいときなどに，can not というように 2 語に分けられることがあるが，cannot のほうが一般的），they have などを用いる．

6.2　化学反応を英語で書く

6.2.1　化学反応の記述法の基本

化学英語に関する英作文で最も重要な表現は，化学反応をいろいろな表現法で記述することであろう．

【例】「**A** と **B** との反応で **C** が生成した」あるいは「**A** は **B** と反応し **C** を与えた」という意味の文章をつくってみよう．

　　The reaction of **A** with **B** produced compound **C**.

　　A reacted with **B** to produce compound **C**.

Point　「**A** と **B** との反応」は，the reaction of **A** and **B** とはせず，通常 **A** with **B** で受ける．また，「**A** を用いる反応」のときは，The reaction using（あるいは employing）**A** とし，The reaction of **A** とは通常しないのが基本．

6.2 化学反応を英語で書く

【練習問題】 次の文を英訳せよ(Chaper 3 の例文 32, 40 などを参照).

グリニャール試薬とアルデヒドとの反応でアルコールが得られる.

→ _____

6.2.2 さまざまな表現をマスターしよう

名詞 reaction と動詞 react を使った基本構文をマスターしよう．次①〜⑤の文章は，内容的にはほぼ同意文である．英語で資料をつくるときには，同じ構文の文章をなるべく続けて用いないように心がけよう．

【例】

① The reaction of **A** with **B** took place (proceeded) at XX ℃ to afford **C** in XX% yield.

　A と **B** との反応は XX ℃で進行し，**C** を XX ％の収率で与えた．

② The reaction of **A** with **B** at XX ℃ afforded **C** in XX% yield.

　A と **B** との XX ℃での反応により，**C** を XX ％の収率で生じた．

③ **A** reacted with **B** at XX ℃ to afford **C** in XX% yield.

　A は **B** と XX ℃で反応し，**C** を XX ％の収率で生じた．

④ **A** was allowed to react with **B** to afford **C** in XX% yield at XX ℃．

　A を **B** と反応させたところ，**C** が XX ℃で XX ％生成した．

⑤ The reaction of **A** with **B** was carried out (performed) at XX ℃ to afford **C** in XX% yield.

　A と **B** との反応を XX ℃で行ったところ **C** が XX ％生成した．

Point 収率を表す％はスペースを入れずに XX％とし，それ以外の単位は XX ℃，XX mol，XX h などとスペースを入れることが多い．

【練習問題1】有機化学の教科書に記載されている化学反応を五つ選び，異なる構文で二つずつ英作文せよ．ただし，「生成した」を意味する単語はすべて異なるものを用いること．

give, produce, prov̲ide, aff̲ord, y̲ield, f̲urnish, g̲enerate, form, result in the formati̲on of, obta̲in

① –1

_____

① –2

_____

② –1

_____

② –2

_____

③ –1

_____

③ –2

_____

④ –1

_____

④ –2

_____

⑤ –1

_____

⑤ –2

_____

続いて，react, reaction 以外の単語を使った表現もマスターしよう．

【練習問題2】 それぞれの構文を用いて，英文を自作せよ．
① **A** is formed by 〜．
② A similar treatment resulted in the formation of 〜．
③ Compound **A** was isolated by 〜．

①

②

③

6.3　辞書を使って表現に注意して書く

自分で実際に論文を書いてみると実感できることであるが，はじめのうちはなかなかスラスラと英文がでてこない．将来，論文を書く機会などがあるような場合には，論文を読むときに，使えそうな表現をピックアップして書き留めておく習慣をつけよう．

【例1】 標準試料を使ってその不斉反応が達成できた．

→ The asymmetric reaction was attained using a standard sample.

→ The asymmetric reaction was achieved when a standard sample was employed.

Point 和英辞典で「標準の」を引くと standard, normal, regular などがあるが，この場合，語感が異なる．リーダーズで normal を引くと「(化学) 1 規定の」の意味があり，誤解を招きかねない．regular は「規則的」という意味がありしっくりこない．

Point when はコンマなしで用いられ「〜したとき」という意味（制限用法）．ここで 〜, when 〜 とすると「そしてそのとき」という意味（非制限用法）になってしまう．

【例2】明らかに彼らの実験データは，われわれのものと一致している．

→ Their experimental data are obviously in accord with ours.

→ Their experimental data are clearly consistent with ours.

Point 和英辞典で「明らかに」を引くと clearly, obviously, evidently, apparently などがヒットする．『化学英語の活用辞典』でも，「明らかに」を見るとこれらの語を使った例文が載せてある．ただし，apparently をリーダーズ英和辞典で引くと，「(実際はともかく) 見たところでは」という意味と，「明らかに」の意味があるので，誤解を避けるためには apparently 以外の語を使用したほうが無難なように思える．次に，「一致する」を『化学英語の活用辞典』で引くと，agree with, be in accord with, be consistent with などがヒットする．

【例3】かさ高い置換基を導入することが，決定的に重要である．

→ It is crucial to introduce a bulky substituent.

→ Introduction of bulky substituent is of crucial importance.

Point 和英では，「重要な」は important がこの意味の最も一般的な語となっている．「決定的に重要な」very important でもよいが，essential や crucial も考えられる (crucial は essential を類語辞典で引くとヒットする)．

【例4】その架橋錯体が触媒となり，発熱反応を促進する．

→ The bridging complex catalyzed the exothermic reaction.

Point この文章は，そのままの語順で英訳してはいけない例である．まず，「その架橋錯体が発熱反応を触媒する」と読み替えればスッキリした英文が書ける．

> **【例5】** 試薬 **A** と試薬 **B** の反応では，化合物 **C** と **D** がそれぞれ 50%，20% の収率で得られた．
>
> → The reaction of **A** with **B** afforded **C** and **D** in 50% and 20% yields, respectively.

Point Chapter 6.2 で学習した化学反応の記述の最も基本形である．書き手および読み手が知っている情報であれば the reaction of A with B ～ とする．

6.4 実践トレーニング 英作文 100

以下の英作文を実際に書いてみよう．

Lesson 1

(1) 遠心分離により未精製の反応混合物から不溶な固体を取り除いたのち，溶媒を濃縮し，^{1}H NMR スペクトルで反応を解析した．
Point 能動態（日本語）→受動態（英語）に変換しよう．
Point 「遠心分離により」は，by centrifuge（遠心分離器）より by centrifugation（遠心分離という手法）のほうがよい．

> Chapter 3 例文 55 と同じく，「機器で解析する」ではなく「手法で解析する」と表すほうがよい．
> Jeff's advice

(2) その実験事実から，その固体は伝導性があると結論づけることができる．
Point 主語を何にするかを考えよう．

(3) その化合物の酸素上の非結合性軌道の電子対が，金属イオンと相互作用することで，求核攻撃に対する活性化が起こる．
Point 「電子対」を修飾する部分が長いことに注意．

(4) 光合成では光化学反応により，本来不可能な吸熱反応が達成される．

(5) 分子およびその集合体の性質は，それらの三次元的構造に大きく依存する．

(6) 各原子は正に帯電した核と，そのまわりを雲状に囲む負に帯電した粒子から成り立っている．
　Point　「～から成り立つ」は，Chapter 3 の例文 96 を参照．

(7) 正に帯電した核子の構成成分は陽子と呼ばれ，その電荷の大きさは負に帯電した電子と呼ばれる粒子の電荷の大きさに等しい．

(8) その混合物をしばらく攪拌したのち，不溶の固体を分離した．
　Point　Chapter 3 の例文 8 を参照．

(9) その化合物の不均一開裂によりカルボカチオンが生成した．
　Point　「生成した」はいろいろな動詞で表せる．

(10) われわれは，その問題に取り組む抜本的な解決方法を必要としている．

Lesson 2

(11) 今日，私はその結晶の充填効率について議論したい．

(12) 紫外線だけでなく可視光も電磁波である．
　Point　as well as を使う場合の主語の人称については，Chapter 3 の例文 98 を参照．

(13) 触媒反応と化学量論反応の違いについて記述せよ．
　Point「〜について記述せよ」にあたる動詞が他動詞ならば，目的語が直接すぐ後ろに来る．

(14) X線は赤外線よりずっと波長が短く，ずっと振動数が高い．

> X線であれ，UVであれ，可視光であれ，赤外線であれ，すべて光は電磁波 electromagnetic wave である．波長が短いほど，波数が大きくエネルギーも大きい．

(15) われわれの提案に対する確固たる証拠を得るために，その反応を検討した．

(16) われわれは，オゾン分解のメカニズムについて研究した．
　Point 読み手がわかっているオゾン分解であれば the ozonolysis とし，一般のオゾン分解のときは抽象名詞扱いで無冠詞．
　Point「〜に関する研究」は，名詞では study on 〜 というが，動詞では直接目的語をとる．

(17) その気体は，加圧下でのみ液化する．
　Point「液化する」をどう表現するか考えよう．
　Point「加圧下」は，under pressure である．under pressurized atmosphere とはいわない．

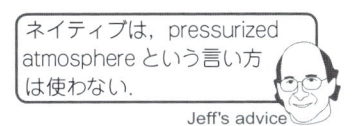

> ネイティブは，pressurized atmosphere という言い方は使わない．
> Jeff's advice

(18) その受容体の NMR スペクトルを測定した．

(19) その結果は，アクセプターの加溶媒分解が起こったことを示唆している．

(20) その触媒反応の主生成物はフェノールであった．

Lesson 3

(21) この化合物のねじれエネルギーは，見積もることさえ難しい．
　Point 「～することさえ難しい」をどう表現するか．

(22) その反応機構は可能性があるが見込みは低い．

(23) これらの抗生物質の組成を GC-MASS で分析した．
　Point 「GC-MASS で」は，by a GC-MASS spectrometer
　　　　よりも，by GC-MASS spectrometry のほうがよい．

ここでの spectrometry は手段を表すので無冠詞．
Jeff's advice

(24) 私の専攻は化学です．

(25) そんな危ないことはしないほうがよい．

(26) これら二つの証拠は，その小さな球形粒子が一つの金原子であることを示している．
　Point 「二つの証拠」が可算名詞か不可算名詞かに注意．

(27) その主題について，体系的な研究はほとんど報告されていない．

(28) フランは脂肪族ジエンと同じくらい付加反応を受けやすい．

(29) 空孔への電子移動が自発的に起こった．

(30) その反応で，分岐鎖をもつすべての原料が消費された．

Lesson 4

(31) 基底状態と遷移状態とのエネルギー差は活性化エネルギーと呼ばれる．

(32) その系はエントロピーを制御することで可逆となる．

(33) 水を水素と酸素に変換する新しい方法を開発した．

(34) その光化学反応を実行するとアンモニアが生成した．
　Point アンモニアは物質名詞なので不定冠詞はつかない．

(35) 水は水素結合のため，沸点が高い．

(36) そのメトキシ基の共鳴効果は誘起効果に匹敵する．

(37) エーテルという名前は，一つの酸素原子に連結した二つのアルキル基をもついかなる化合物をも指す．
　Point 「〜を指す」は Chapter 3 の例文 1 を参照．

(38) 今回と前回の実験結果に基づいて，官能基選択的な新規反応を開発したい．

136 ● Chapter 6　化学を英文で書き表そう

(39) 3か月というのは，その実験を終えるには短すぎる．

(40) S_N2 反応の立体化学は通常，反転である．

Lesson 5

(41) ラジカルの再結合で連鎖反応が終了する．

(42) その試料は容易に励起され，ただちに可視光を放射する．

(43) カップリング定数の大きさからその化合物の溶液中でのコンフォメーションが推測できる．

(44) 光照射は，反応速度に大きく影響をおよぼす．
　Point Chapter 3 の例文 66 を参照．

(45) 伸縮性のある伝導性ポリマーを合成した．

(46) この化合物の合成は，次の三つの方法いずれかにより達成できる．

(47) この化合物の合成は，次の二つの方法いずれかにより達成できる．

(48) この論文によると高い光学収率が達成される．

(49) IR の指紋領域はそれぞれの化合物に特有である．

(50) 質量分析法によって物質の重さが確かめられる．

Lesson 6

(51) 不斉増幅とは，用いる触媒より高い e.e. の生成物を生じる現象である．
Point 「現象」にかかる修飾部分が長いので，関係代名詞を用いる．

(52) 一方，不斉増殖とは，少数の不斉分子から多くの不斉化合物を合成する反応のことである．

(53) ボスはわれわれにもっとよい方法論を提案するよう指示した．
Point 直接の指導教官は boss と呼ばれることが多い．

(54) 触媒的な窒素の取り込みは，今でも化学者にとって困難な課題である．

(55) 系中で発生する水が触媒を不活性にする．
Point Chapter 3 の例文 53, 108 を参照．

(56) カリウムカチオンはその化合物を受容体とは認識しない．

(57) そのラジカルは一重項でも三重項でも存在しうる．

(58) キラル配位子の利用が不斉合成に革新をもたらした．
　Point 「利用する」「使用する」は，use と employ がよく使われる．

(59) ESR スペクトルの測定には石英でできたセルを使用する必要がある．
　Point 能動態（日本語）→受動態（英語）への変換．

(60) 抗体は病気と闘うためにわれわれの体内でつくられる物質である．

Lesson 7

(61) 合金とは，二つあるいはそれ以上の金属が混ぜ合わされたものである．

(62) 硫黄が架橋した標準試料を作製した．
　Point 「標準試料」の「標準」に注意．

(63) その多孔性ゼオライトは，少量の水を含んでいる．

(64) そのゼオライトを酸で処理すると触媒が再生した．
　Point 「処理すると～が再生した」を SVO 構文で表すとどうなるか．

(65) 彼は分子カプセルの定義について説明した．
　Point 「説明する」にあたる動詞が他動詞ならば，すぐ後ろに目的語が来る．

(66) 反応容器を炉から取りだし空冷した．
　Point 能動態（日本語）→受動態（英語）への変換．

(67) アルキル基の電子的効果は，超共役が関与していることを示唆した．

(68) 電子吸引性基があるとメタ配向性を示す．
　Point　「〜があると」は，「〜が存在することは」と言い換えられる．

(69) 優先的に一つの異性体のみが生成する理由ははっきりしない．

(70) 高いひずみエネルギーのため，分解が容易に起こる．

Lesson 8

(71) 君は，その化合物の同定のために標準試料を合成する必要がある．

(72) 周期表の左の元素は電気的に陽性である．

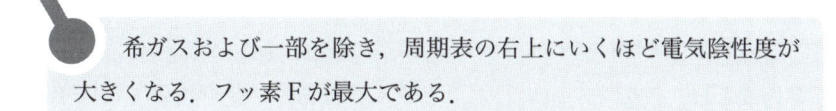

> 希ガスおよび一部を除き，周期表の右上にいくほど電気陰性度が大きくなる．フッ素Fが最大である．

(73) われわれは，くどい表現のない論文を書く必要がある．

(74) 再現性から判断すると，この方法のほうがずっとよい．

(75) この薬は効き目がない．

(76) 絶対零度は熱力学において大きな意義がある．

(77) その結果は，π受容体の特徴から予想されるものと一致する．

(78) ニトリルはカルボン酸の合成等価体である．

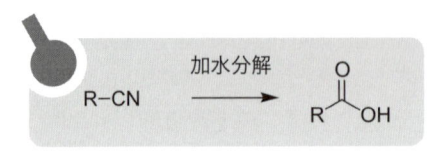

$$R-CN \xrightarrow{加水分解} R-COOH$$

(79) C–C結合切断を伴う開環重合により直線状のポリマーが得られる．
Point 受動態「得られる」を能動態にして考える．

(80) その1,3-双極子の反応性がこれらの実験より確かめられた．

Lesson 9

(81) 絶対不斉合成とは，円偏光や磁場など，キラルではあるが化学物質でない不斉源を用いる反応である．

(82) その結果は，二量体の錯体の生成と一致している．

(83) 彼らは，大気汚染物質の放出を抑制する新規な方法論を明らかにした．

(84) フォトクロミズムは，電磁波の照射によって誘起される異なる吸収スペクトルをもつ二つの化学種間の可逆的な変化と定義される．

(85) 水の電気分解により陰極では水素が発生する．
　Point 「陰極では」で用いる前置詞に注意．

(86) その巨大分子の安定化には分子内のπ–πスタッキングが決定的な役割を果たしている．
　Point Chapter 3 の例文 125 を参照．

(87) A 社は，そのネマチックポリマーが液晶ディスプレイの素材として使用できることを実証した．

(88) その論文は，信じがたいが信頼性のある事実について言及している．

(89) 彼のその危険に対する認識は不十分であった．

(90) この注射器は使い捨てです．

Lesson 10

(91) 彼女は博士号を取得するための必要条件をすべて満たした．

(92) エーテルは引火性で有毒な液体である．

(93) 彼らは試薬びんのリサイクルの問題に取り組んだ．

(94) 常磁性であるという性質は，その構造の可能性を否定するものである．

(95) その試薬は多孔性ゼオライトの中に拡散する．

(96) レンズの真ん中より両端が厚いと光線を発散させる．
　　Point　a lens を主語にしたときに「真ん中より両端が厚い」をどう表現するか．

(97) その二つの仮定はどちらも正しくない．

(98) おのおのの原子軌道は量子化されており，それぞれの元素固有のエネルギーをもつ．

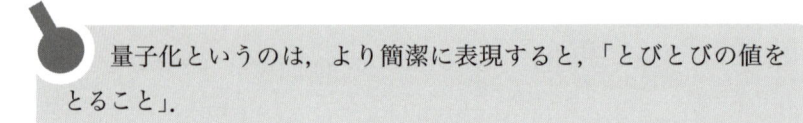

> 量子化というのは，より簡潔に表現すると，「とびとびの値をとること」．

(99) 遷移状態と中間体はともにエネルギー微分がゼロである．

(100) ラジカルとは SOMO をもつ化学種のことである．

> SOMO: singly occupied MO あるいは semi-occupied MO．

Chapter 7 大学院入試問題に挑戦しよう

> これまでの学習の総決算として,大学院入学試験問題を解いてみよう.単に大学院入試対策に利用するだけでなく,じっくりと時間をかけて解答することで,確実な英語力アップにつながるはずである.

7.1 大学院入試問題の傾向と対策

　このChapterを執筆するにあたって,十数校の大学院入学試験問題を取り寄せて解答してみた.いくつか気づいたことがあるので,受験を控えている人は参考にしていただきたい.

① 同じ試験でも,設問により難易度にかなりの差がある場合がある
　　たとえば,2問の長文読解問題が出題されている場合,どちらが簡単そうかまず見きわめたほうがよさそうだ.これは,難易度に差をつけて実力の差を正確に判定したいという出題者の意図による場合や,出題者が複数の場合に起こりうる.当然のことながら,簡単と判断される問題から解答していこう.

② 過去の出題の傾向をつかんでおく
　　出題者は毎年交代している可能性が高いが,問題全体の構成・配点は毎年同じことがほとんどである.

③ できるだけ多くの問題に手をつける
　　とくに英作文は,問題が長文問題のあとにある場合が多いが,解答すれば部分点がもらえると考えられる.長文問題に取りかかる前に,ひと通り解答するのも一つの方法だろう.不完全でもざっと英文が書けるよう練習しておきたい.

④ 長文問題は,まず設問に目を通す
　　どのようなテーマについて書かれた英文なのか,何について答えればよいのかわかるはずだ.また,長文中の英文和訳など,必ずしも全文の内容を理解できていなくても解答できてしまうような問題もある.時間を有効に使おう.

7.2 実践トレーニング 入試問題 7

Lesson 1 次の英文の下線部を和訳せよ．

(a) Nanotechnology refers broadly to using materials and structures with nanoscale dimensions, usually ranging from 1 to 100 nanometers (nm). Perhaps, without realizing it, we already encounter some likeness of nanotechnology in daily life. For example, proteinaceous molecules in living organisms from bacteria to beetles to humans serve as "molecular motors" to drive everything from flagella motion to muscle flexion. Nanometer-sized particles have been developed to improve the mechanical properties of tires, initiate photographic film development, and serve as vital catalysts in the petrochemical industry.

(b) Nanotechnology is not just about the size of very small things. More important, it is about structure and the ability to work at the atomic or molecular level. This results in materials and systems that often exhibit novel and significantly changed physical, chemical, and biological properties due to their size and structure. These new properties include improved catalysis, tunable wavelength sensing ability, and increased mechanical strength.

The basic structures of nanotechnology include nanoparticles or nanocrystals, nanolayers, and nanotubes. (c) These nanostructures differ in how they are made and how their atoms and molecules are ordered. A nanoparticle —— a collection of tens of thousands of atoms measuring about 1-100 nm in aggregate diameter —— is the most basic structure in nanotechnology. Such nanoparticles are created atom by atom, so the size and often the shape of a particle are controlled by experimental conditions. These particles can also be described as nanocrystals because the atoms within the particle are highly ordered, or crystalline.

(d) Nanostructures are often arranged or self-assembled into highly ordered layers arising from hydrogen bonding, dipolar forces, hydrophilic or hydrophobic interactions, gravity, and other forces. Many naturally occurring biological structures, like membrane, vesicles, and DNA form because of such self-assembly. Repeating structures with a tailored periodicity are essential in many applications of nanotechnology, such as photonics, catalysts, and membranes. Understanding and building nanostructures through self-assembly is the core of the nanotechnology creation process.

［大阪大学大学院工学研究科博士前期課程物質化学専攻　平成 16 年度］

> **Keywords**
>
> refer to 〜　　　　　〜について言及する　　differ in + that 節　〜の点で異なる
> more important = more importantly　　　　（ここでは that 節の代わりに how 〜）
> result in 〜　　　　　〜に帰着する　　　　　arise form 〜　　　〜から生じる
> due to 〜　　　　　　〜のためで

Lesson 2　次の英文を読み，内容を 300 字程度に要約せよ．

At the Goethe Institute in Mannheim,* I studied German Language for eight weeks, starting in the beginning of August 1988. Later on, I joined the research group of Professor Heinrich Wieland of Heidelberg University. While I was there, my wife and one-year-old daughter came and stayed together with me. I enjoyed doing the research work there. Heidelberg University is one of the most famous and oldest universities in Germany. The two years I spent there have been fruitful, both academically and personally. Our family had the great opportunity to learn about people in Germany and their customs and traditions. That was a valuable and rewarding experience both for my wife and me.

My research project in Heidelberg included the synthesis and structural characterization of novel polynuclear coordination compounds with ruthenium polypyridine unit and are of particular importance due to the photophysical properties of ruthenium[II]. I would like to especially mention our novel coordination compounds that were successfully synthesized during my stay in Heidelberg. I have carried out my research project successfully with the provisions of good research facilities. I worked together with two ambitious German Ph.D. students. I have gained many skills and much knowledge from them. I hope that they were also able to learn something from me. The experience and knowledge gained in Heidelberg is of considerable assistance to further developments in higher education and research in my own country. It is very helpful in giving lectures, supervising the research of post-graduate students, measuring

new data with our instruments and interpreting the measured data. In future, I intend to continue my current research as a university teacher.
*Goethe Institute in Mannheim（ライン河畔の町マンハイムにあるドイツ語学校）

［大阪大学大学院工学研究科博士前期課程物質化学専攻　平成 16 年度］

Point　単語・構文ともに平易だが，要約問題では，内容をまとめる日本語能力も必要．

Lesson 3　次の英文を読み，設問(1)〜(2)に答えよ．

Chemistry! We all associate chemistry with test tubes, stinking laboratories and explosions. (a) Perhaps the development of new knowledge in chemistry, more than any other science, has been characterized as a sparkling interplay between theory on one hand, the safe and predictable, and on the other hand, the explosive and surprising reality. When we by chance discover something that may become valuable, we talk about (b) "serendipity"　− after the tale about *The Three Princes of Serendip*, who traveled widely and had the gift of drawing far-reaching conclusions from whatever they encountered.

Let us go back to the beginning. In Japan, in 1967, a group of scientists were studying the polymerization of acetylene into plastics　− acetylene was the gas that the Swedish engineer Gustaf Dalen once tamed to bring light in the dark for sailors in the form of blinking buoys. Polymerization is the process by which many small molecules react to form a long chain − a polymer. Professors Ziegler and Natta were awarded the Nobel Prize for a technique for polymerizing ethylene or propylene into plastics; the Japanese scientists used the same catalyst for polymerizing acetylene. One day a visiting researcher in the laboratory, the story goes, added more catalyst than written in the recipe: actually one thousand times too much! Imagine the surprise among your invited dinner guests if, rather than using a few drops of Tabasco in the soup, you had added the whole bottle! The result was a surprise also to the scientists. Instead of the expected black polyacetylene powder that normally was obtained, and that was of no use, a beautifully lustrous silver colored film resulted. It was, however, only its appearance that was metallic. The material did not conduct electricity. The breakthrough was not made until ten years later in collaboration between physicist and chemists, continuing the experiments with the silver colored film.

They tried to oxidize the film using iodine vapor, and － Bingo! The conductivity of the plastic increased by as much as ten million-fold; it had become conductive like a metal, comparable to copper. This was a surprising discovery, to the researchers as well as to others － we are all used to plastics, in contrast to metals, being insulators, which is why we cover electrical cords in plastic.

The exiting idea of being able to combine the flexibility and low weight of plastics with the electric properties of metals has stimulated scientists all over the world, resulting in a novel research field bordering physics and chemistry. Various theoretical models and new conductive, but also, semi-conductive, polymers followed during the 1980s in the wake of the first discoveries. Today we can see several possible applications. How about electrically luminous plastic that may be used for manufacturing mobile phone displays or the flat television screens of the future? Or the opposite － instead using light to generate electric current: solar-cell plastic that can be unfolded over large areas to produce environmentally friendly electricity. Finally, lightweight rechargeable batteries may be necessary if we are to replace the combustion engines in today's cars with environmentally friendly electric motors － another application where electrical polymers might find use. In parallel with development of conducting polymers, there is an ongoing development of what we might call "molecular electronics," where the very molecules perform the same tasks as the integrated circuits we just heard about in Physics, with the difference that these could be made incomparably smaller. In laboratories around the would, scientists are working hard to develop molecules for future electronics. And among test tubes and flasks, and in the interplay between theory and experiment, we may some day again be astonished by something unexpected and fantastic.

(1) 下線(a)を和訳せよ．

(2) 下線部(b)について次の設問に答えよ．
 (i) この文章における"serendipity"とはどういう意味か？ 20字以内で述べよ．
 (ii) 筆者が童話"*The Three Princes of Serendip*"になぞらえている内容を200字程度で要約せよ．

［大阪大学大学院工学研究科博士前期課程物質化学専攻　平成16年度］

Keywords

associate A with B	AでBを思いだす，AとBとを結びつけて考える	lustrous	光沢のある，輝く
stinking	悪臭のある	in collaboration between A and B	AとBとの共同製作により
by chance	偶然に	be comparable to ～	～に匹敵する
gift	才能	luminous	発光する
far-reaching	（影響などが）遠くまで及ぶ	electronics	電子機器
tame	利用できるように制御する	the very molecules	まさしくその分子
blinking	明滅する	integrated circuit	集積回路
buoy	ブイ，浮標	incomparably	比較にならないほど
recipe	レシピ		

Lesson4 次の文章を読んで，以下の設問 (1)～(3) に答えよ．

　　Between 3.2 and 1.8 billion years ago there was little or no dioxygen in the atmosphere and, therefore, no protective ozone layer. No life was possible on the dry land because of the intense ultra-violet radiation. Two processes were responsible for the production of dioxygen. The high-energy band of intense ultra-violet radiation (UVC) from the Sun caused the photochemical decomposition of water molecules with the production of dihydrogen and dioxygen. The dihydrogen gas was molecules escaped the Earth's gravity so that the dioxygen content of the atmosphere increased. It has been estimated that when the amount of dioxygen in the early atmosphere reached only 0.1% of its present level, the photolysis of water was prevented from occurring by the dioxygen itself preferentially absorbing the ultra-violet radiation. A more significant source of atmospheric dioxygen was life in the form of the first bacterial cells which were able to use visible light to convert carbon dioxide into carbohydrates and dioxygen; the life-producing process of photosynthesis. It is thought that such early life forms were active around 3.2 billion years ago. The photosynthetic bacteria caused the dioxygen content of the atmosphere to increase at the expense of the carbon dioxide content until it reached a level of 1% of the present value. At such a concentration it became possible for organisms to exist which depended upon respiration for their energy requirements. Examination of the fissile record indicates that the dioxygen level of 1% of the present value was achieved around 2 billion years ago. The ozone layer had not fully developed so that life was still confined to shallow water which offered protection against the ultra-violet radiation which was incident upon the Earth's surface at the time.

Respiration reverses the photosynthetic process, converting dioxygen into carbon dioxide, but the rapidly multiplying photosynthetic organisms produced sufficient dioxygen to cause its atmospheric concentration to increase. (i) When the level had build up to 10% of the present day value, the production of the ozone layer in the upper atmosphere was sufficiently developed to protect the Earth's surface from the more damaging portion of the ultra-violet radiation (UVC). This important stage was reached around 400 million years ago, so that life could begin to evolve on the land from that time. Early vegetation developed and contributed to the formation of the deposits of coal between 360 and 290 million years ago. Some of the oxygen was consumed in the formation of oxides in the outermost section of the Earth's crust but sufficient dioxygen built up in the atmosphere to sustain the protective ozone layer and so prepare the Earth for the subsequent development of its various extra life forms. (ii) The development of vegetation reduced the proportion of carbon dioxide and increased that of dioxygen as photosynthesis proceeded, so that the main constituents of the atmosphere become dioxygen and dinitrogen with a small proportion of carbon dioxide.

設問(1) 上記文章に記載されている大気中の酸素分子の生成過程二つを，それぞれ100文字以内で説明せよ．

   (a) 1番目の生成過程： (b) 2番目の生成過程：

設問(2) 本文中にイタリック体で示した"*respiration*"の意味を10文字以内で記せ．

設問(3) 本文中に下線を付した(i)および(ii)の文章を和訳せよ．

[大阪大学大学院工学研究科博士前期課程物質化学専攻　平成16年度]

Keywords

UVC	ultraviolet radiation C の略. C 波紫外線	multiply	繁殖する
		vegetation	草木
preferentially	優先的に	outermost	最も外側の
at the expence of ～	～という犠牲を払って	crust	地殻
organism	生物体，（微）生物	build up	蓄積する〔自動詞〕
be incident upon ～	（光線が）～に投射する		

Point
- 横軸に時間を取った時系図を作成しながら読むと理解しやすい．
- 1 行目の「18〜32 億年前には，ほとんどあるいはまったく酸素がなく，オゾン層は存在しなかった．」という記述は第 1 パラグラフ全体にかかっている．したがって，18〜19 行目の「20 億年前の酸素レベル（1％）」は，O_2 がほとんどない年代に属している．

Lesson 5 次の和文の下線部を英訳せよ．

(a) 気体の性質は，その圧力と温度によって特徴づけられる．また，気体分子の集団が持つ性質である圧力や温度はそれぞれの分子の力学的運動と関係づけられる．容器中の気体分子はたえず無秩序な運動をし，他の分子や器壁との衝突を繰りかえしている．いま，気体中に分子の速度に注目すると，それぞれの分子はさまざまな速度をもって運動している．また，個々の分子も衝突により速度を変化させている．(b) しかし，平衡状態では，多数の分子の速度の平均値は変化しないと考えることができる．このような場合には，分子の集団としての気体の性質には個々の分子の運動の個性は反映せず，平均化された姿をあらわす．

分子の運動は振動，回転，並進の運動からなりたっている．簡単のため，分子間の衝突は弾性的であるとする．(c) すなわち，分子の並進運動のエネルギーは，衝突に際して電子や振動，回転のエネルギーに変わることはないと考える．さらに，分子間ポテンシャルは剛体ポテンシャルだとする．以上の条件は，分子をビリヤードの玉に似た完全弾性体とみなすことによって満足される．また，分子の大きさは容器にくらべて非常に小さい．(d) 希薄な気体では，容器中の分子が占める体積は容器の体積にくらべて無視することができる．以上の条件は，気体が理想気体としてふるまうための条件である．

［大阪大学大学院工学研究科博士前期課程物質化学専攻　平成 16 年度］

> **Point**
> ・「特徴づける」は characterize．
> ・S consider O to be 〜　S（人）がO（人・物）を〜だとみなす（例文 35）．
> ・「物体の速度」は velocity．
> ・変化しない→一定である＝ constant も可．
> ・「すなわち」は in other words も可．ただし，namely は普通，挿入的に用い，文頭には用いない．collide は，with 〜 などの副詞句を伴う．解答例では，with each other を補った．
> ・negligible は「可能である」という意味を内在しているので，can be negligible とはしない．

Lesson 6　次の和文(1)，(2)を英訳せよ．

(1) ある原子の第1イオン化エネルギーとは，基底状態にある1モルの気体状原子を，1モルの気体状1価陽イオンに変換するために必要な最小エネルギーである．一方，電子親和力とは，1モルの気体状原子に電子が付加することにより，1モルの気体状陰イオンが生じる場合の内部エネルギーの変化として定義される．

(2) カルボカチオンは，アルケン，エノールエーテル，エナミン，芳香族化合物のような電子豊富な化合物と炭素–炭素結合の生成を伴って反応する．合成上で有用なカルボカチオンは，その酸化レベルに応じて3つのクラスに分類できる．第一が，ハロゲン化アルキルやアルコールの不均一分解により得られるアルキルカルボカチオンである．第二のグループはアルデヒドやケトン由来のものである．最後のグループは Lewis 酸触媒存在下での酸誘導体の結合開裂により得られるアシリウムイオンである．

［大阪大学大学院工学研究科博士前期課程物質化学専攻　平成 16 年度］

> **Point**
> ・(2) の第1文は，基本構文 A reacts with B to produce (form, afford など) 〜 で記述可能．
> ・「〜に分類される」は Chapter 3 の例文 113 を参照．

Lesson 7 以下の文章 (Freeman Dyson 著, "IMAGINED WORLD" の一節) をよく読み，(1)～(3) に答えよ．

If we are looking for new direction in science, we must look for scientific revolutions. When no scientific revolution is under way, science continues to move ahead along old directions. It is impossible to predict scientific revolutions, but it may sometimes be possible to image a revolution before it happens.

There are two kinds of scientific revolutions, those driven by new tools and those driven by new concepts. Thomas Kuhn in his famous book, *The Structure of Scientific Revolutions*, talked almost exclusively about concepts and hardly at all about tools. His idea of a scientific revolution is based on a single example, the revolution in theoretical physics that occurred in the 1920s with the advent of quantum mechanics. This was a prime example of a concept-driven revolution. (a) <u>Kuhn's book was so brilliantly written that it become an instant classic.</u> It misled a whole generation of students and historians of science into believing that all scientific revolutions are concept-driven. The concept-driven revolutions are the ones that attract the most attention and have the greatest impact on the public awareness of science, but in fact they are comparatively rare. In the last 500 years, in addition to the quantum-mechanical revolution that Kuhn took as his model, we had six major concept-driven revolutions, associated with the names of Copernicus, Newton, Darwin, Maxwell, Freud, and Einstein. During the same period there have been about twenty tool-driven revolutions, not so impressive to the general public but of equal importance to the progress of science. Two prime examples of tool-driven revolutions are the Galilean revolution resulting from the use of the telescope in astronomy, and the Click-Watson revolution resulting from the use of X-ray diffraction to determine the structure of big molecules in biology.

The effect of a concept-driven revolution is to explain old things in new ways. The effect of a tool-driven revolution is to discover new things that have to be explained. In almost every branch of science, and especially in biology and astronomy, there has been a preponderance of tool-driven revolutions. We have been more successful in discovering new things than in explaining old ones. In recent times my own field of physics has had great success in creating new tools that have started revolutions in biology and astronomy. (b) <u>Physics has been less successful in creating new concepts with which to understand its own discoveries.</u>

(1) 筆者は科学革命にはどのような種類のものがあり，それらの科学革命にはどのような相違点があると考えているのか．200字程度の日本語で説明せよ．

(2) Thomas Kuhn は科学革命についてどのように考えているか．筆者の考え方との相違がわかるように日本語で簡潔に述べよ．

(3) 下線部(a)，(b)を和訳せよ．

［京都大学大学院理学研究科化学専攻修士課程入学試験 平成16年度］

Keywords

quantum mechanics	量子力学	astronomy	天文学
Galilean	ガリラヤ人の	preponderance	優勢

実践トレーニングの解答例

Chapter 5

Lesson 1

新規な不斉アシル化触媒を用いる反応を開発した．生成物は，容易にラセミアルコール **A** の速度論的分割によって分割でき，ほとんどの場合に高いエナンチオ選択性が達成できる．結果は，触媒と基質の間での水素結合相互作用が，この触媒的アシル化反応における高いエナンチオ選択性に起因することを示唆している．反応機構を解明するため，溶媒依存性に関してさらに研究を行った．反応は，異なる置換基をもつほかのさまざまなアルコールにも適用可能である．研究したなかでは，エタノールアミン誘導体が，とくに高い有効性(up to S = 16.8)を示した．

Lesson 2

RNA が遺伝情報の受け渡しに関与しているだけでなく，触媒能をもつかもしれないということは，1980 年代の初めから知られていた．1989 年にノーベル賞の栄誉が与えられたこの発見は，遺伝情報の貯蔵と化学反応の制御がともに RNA によって実行されるという "RNA ワールド" が存在するのではないかという初期の仮説に触発されたものである．近年では，試験管内での選択的な発生の技術が，RNA の触媒としての潜在能力の探求に大きく貢献した．RNA の触媒としての潜在能力の探求は，初期には，アシル化やアルキル化のような RNA に存在する官能基での反応に照準が合わされていたが，最近の研究では，人工的に DNA に連結したジエンを用いて Diels-Alder 反応のような幅広い反応が可能であることを示している．

Lesson 3

それらの属名にもかかわらず，希土類は稀少でもなければ土の成分(金属酸化物)でもない．それらは金属元素であり，一つを除いては金，銀，水銀，タングステンなどより豊富に地球に存在する．つまり，希土類元素は，実際には遍在し，実質上すべての鉱物に低濃度ながら含まれている．しかし，一般にある鉱物から希土類を抽出するのは費用がかかりすぎる．それゆえ，元素を経済的に抽出するのに見合う十分な濃度をもつ，ひと握りのあまりよく知られていない鉱物が実際に存在することは幸運なことである．なぜなら，希土類元素は，現代の数多い製品の製造における鍵となる成分であるからである．セリウムとエルビウムは高性能合金の構成要素である．ホルミウム，ジスプロシウムは，ある種のレーザー結晶に必要である．ネオジムは，今までに知られている最も強力な永久磁石の鍵となる成分であり，それは新しいモーターの設計を可能にする．おそらく，希土類元素の最もおもしろい応用は，かなり高い温度でも超伝導となるセラミックを製造することであろう．

Lesson 4

顕微鏡のスイッチをオンにし，光をつける．次に，スライドを顕微鏡のステージに置く．試料に焦点が合うまで対物レンズを下げる．試料がレンズの真ん中にくるように調節する．そして，高倍率レンズを試料がより高倍率で覗けるようにぐるっと回す．測定終了後，スライドを片付け，スイッチを切る．

Lesson 5

"有機化学"という言葉を耳にしたとき，何がみなさんの頭に思い浮かぶだろうか？ ある人は，"生命の化学"あるいは"炭素の化学"と考えるかもしれない．また，なかには，"医学部に進むためのもの"，"重圧"，"難解"，あるいは"暗記もの"という人もいるだろう．形式的には，炭素化合物の学問ではあるが，有機化学という学問は，多くのほかの分野で必要とされる共通の能力を含む．つまり，有機化学は一つの自然科学であるけれども教養科目でもあり，その概念と技術に精通することは，ほかの分野の学問の能力を高めることにもつながりうる．

Lesson 6

本書の初版が出版されてはや6年が経ち，この間，(化学界では)多くの変化があった．遷移金属を有機合成に使うことは，有機合成のコミュニティーでは広く受け入れられるようになり，少なくともいくつかの遷移金属を介する鍵過程を含まない，複雑な化合物の全合成を見つけるのはまれである．加えて産業界においても，均一系触媒を使用することに対する嫌悪感はほとんどなくなった．ひいては，ポリマーや材料の合成，固相でのコンビナトリアル合成においても遷移金属が非常に頻繁に使用されるようになった．

Lesson 7

ASCミーティングから
有機結晶のなかでのC–C結合の形成
固体状態での光化学反応が隣接する第4級中心をつくる新たな方法を提供する

隣接する第4級中心(炭素)は，合成有機化学での最も難しい構造的モチーフのうちの一つである．カリフォルニア大学ロサンゼルス校の化学教授ミゲル・A・ガルシア-ガリベイは，とりわけ高い価値をもつ化合物に対しては，結晶化したケトンの光化学的脱カルボニル化による炭素–炭素結合形成がこのモチーフをつくる見込みのある選択肢の一つだと考えている．先月，ガルシア-ガリベイはニューヨークのACSナショナルミーティングで，この反応の，複雑な分子の全合成への応用について発表した．
　ガルシア-ガリベイは，隣接する第4級キラル中心をもつ天然物を合成することは，溶液化学ではきわめて困難である，と語る．一方，光化学脱カルボニル化に適する基質ができると，固体状態での反応は直接的で，目的とする生成物を最高95%収率，100%に近い鏡像体過剰率で与える．

考え方はシンプルである．結晶化した，六つの異なる置換基をもつキラルなアセトン誘導体をつくり，光化学的にカルボニル基の二つの結合を切断する．COが取り除かれると，生じるラジカルはその後連結し，完全に基質の絶対立体配置を保った結合をつくる．「一般には，ラジカルはキラルではない．しかし，結晶のキラルな環境下では配置的にトラップされてしまう．そして，二つの電子は互いにとても近いので，それらはすぐに望みの結合をつくる」とガルシア-ガリベイは説明する．

Chapter 6

Lesson 1

(1) After the insoluble solid was removed from the crude reaction mixture by centrifugation, the solvent was evaporated and the reaction was analyzed by ^{1}H NMR spectroscopy.

Point 一般的な方法を表す抽象名詞は通常，無冠詞で用いる．by the ^{1}H NMR spectroscopy としない．

(2) ① The experimental data clearly showed that the solid is conductive.

Point has conductivity としても文法的には誤りでないが，is conductive のほうがより直接的なニュアンスがでる．

私なら後者の表現を使う．
Jeff's advice

② We can conclude from the experimental data that the solid is conductive.

(3) ① The interaction between the metal ion and the electron pair, which is in the non-bonding orbital on oxygen, activates the nucleophilic attack.

② The interaction between the metal ion and the electron pair, which is in the non-bonding orbital on oxygen, facilitates the nucleophilic attack.

Point 「電子対」を修飾する語が日本語ではその語の前に置かれているのに対し，英語では関係代名詞を使って修飾するために後に置かれる．そのため，「電子対と金属イオンの相互作用」を「金属イオンと電子対の相互作用」と等価と考え，順番を入れ替えたほうが英訳しやすい．

(4) ① The photochemical reactions of photosyntheses make normally impossible endothermic reactions realizable.

② Normally impossible endothermic reactions can be achieved by the photochemical reactions of photosyntheses.

(5) The nature of a molecule or its assembly significantly depends on three dimensional structure.

(6) Each atom comprises a positively charged nucleus and negatively charged particles. The latter surround the former like cloud.
Point 英語で1文で表現しにくい文は，構文を変えて2文にすることを考える．

(7) The positively charged component in the nucleus is called a proton, whose total charge is equal to that of the negatively charged particles called electrons.

(8) After the mixture was stirred for a while, the insoluble solid was separated.
Point 「不溶の固体が分離した」なら the insoluble solid separated となる．

(9) Heterolysis of the compound produced a carbocation.

(10) We need a radical solution to tackle the problem.

Lesson 2

(11) Today, we wish to discuss the packing coefficient of the crystal.

(12) Visible light as well as UV rays are electromagnetic waves.
〔私なら，この文章では is でなく are を使う．— Jeff's advice〕

(13) Describe the difference between a catalytic reaction and a stoichiometric reaction.

(14) X-ray radiation has a much shorter wavelength and much higher frequency than infrared radiation.

(15) The reaction was examined to get solid evidence for our proposal.

(16) We studied the mechanism of the ozonolysis.

(17) ① The liquefaction of the gas only occurs under pressure.

② The gas liquefies only under pressurization of the atmosphere.

(18) The NMR spectrum of the acceptor was measured.

(19) That result indicates that solvolysis of the acceptor took place.

(20) The major product of the catalytic reaction was phenol.

Lesson 3

(21) It is difficult to even estimate the torsional energy of the compound.

(22) This reaction mechanism is possible but not likely.

(23) The compositions of these antibiotics were analyzed by GC-MASS spectrometry.

(24) ① My major is chemistry.

　　　② I major in chemistry.

(25) It's better not to take the risk.

(26) These two pieces of evidence show that the small spherical particle is a gold atom.
　　Point evidence, damage, equipment, information, progress などは通常，不可算名詞扱い．two evidences としない．

(27) ① Few systematic studies have been reported on the subject.

　　　② Little systematic research has been reported on the subject.
　　Point few は可算名詞，little は不可算名詞に用いる．study は可算名詞，research は不可算名詞扱いされることに注意せよ．

(28) Furan is almost as susceptible to addition as aliphatic dienes.
　　Point furan はある一つの特定の物質名詞，一方脂肪族ジエンは化合物の種類を示す普通名詞．

(29) Electron transfer to a hole has occurred spontaneously.

(30) All starting materials with branched chains were consumed by the reaction.

Lesson 4

(31) The energy gap between the ground state and the transition state is called the activation energy.

(32) The system becomes reversible by controlling the entropy.

(33) A new method for converting water into hydrogen and oxygen has been developed.

(34) When the photochemical reaction was carried out, ammonia was generated.

(35) Water has a high boiling point because of hydrogen bonding.
Point hydrogen bonding は抽象名詞扱い．一方，hydrogen bond は可算名詞扱いで，こちらを使う場合は hydrogen bonds とする．

(36) The resonance effect of the methoxy group is comparable to its inductive effect.
Point 「～に匹敵する」は be comparable to ～ が定番の表現．

(37) The name ether refers to any compound that has two alkyl groups linked through an oxygen atom.

(38) Based on the results of the present and previous experimental data, we want to develop a novel chemoselective reaction.

(39) Three months is too short a period of time to finish the experiment.
Point 全体で一つの単位を示す主語は，単数で受ける．

(40) The stereochemistry of an S_N2 reaction is usually inversion.

Lesson 5

(41) Radical recombination terminates a chain reaction.

(42) The sample is easily excited and immediately emits visible light.

(43) We can infer the conformation of the compound in solution from the coupling constant.

(44) ① Photoirradiation significantly affects the reaction rate.

② Photoirradiation significantly influences the reaction rate.

(45) An electrically conducting polymer with elasticity has been synthesized.

(46) The synthesis of this compound can be performed by any of the following three methods.

(47) The synthesis of this compound can be performed by either of the following two methods.
Point either は，肯定文では「どちらの〜でも」，否定文では「どちらの〜も」の意．I don't know either method. は「どちらの方法も知らない」となる．なお，二つのときは either を用い，三つ以上のときは any を用いる．

(48) According to this paper, a high optical yield is achieved.

(49) The fingerprint region of the IR spectra is unique to each compound.
Point each は単数可算名詞につく．

(50) The weight of a substance is confirmed by mass spectrometry.

Lesson 6

(51) Asymmetric amplification is a phenomenon that produces higher e.e. in the product than in that of the catalyst employed.

(52) On the other hand, asymmetric multiplication is a reaction that produces many asymmetric compounds from a small number of asymmetric molecules.

(53) The boss instructed us to come up with much better methodology.

(54) Catalytic incorporation of dinitrogen is still a challenging subject among chemists.

(55) The water generated *in situ* inactivates the catalyst.

(56) A potassium cation does not recognize the compound as a receptor.

(57) That radical can be either in a singlet or triplet state.

(58) Employment of chiral ligands brought about innovation in asymmetric syntheses.

(59) A cell made of quartz must be used for measurement of the ESR spectrum.

(60) An antibody is a substrate produced by our bodies to fight disease.

Lesson 7

(61) An alloy consists of two or more metals mixed together.

(62) An authentic sample with bridged sulfur was prepared.

(63) That porous zeolite contains a small amount of water.

(64) Treatment of the zeolite with the acid regenerated the catalyst.

(65) He explained the definition of molecular capsule.

(66) The reaction vessel was taken out of the furnace and cooled in air.

(67) The electronic effect of the alkyl group suggested that hyperconjugation was participating.

(68) The presence of an electron-withdrawing group results in a meta-orientation.

(69) The reason why predominantly only one isomer is produced remains unclear.

(70) Decomposition takes place easily due to the high strain energy.

Lesson 8

(71) You should prepare an authentic sample to identify the compound.

(72) Elements seen on the left side of the periodic table tend to be electrically positive.

(73) We need to write a paper without redundant expressions.

(74) Judging from the reproducibility, this method is much better.

(75) This medicine has no efficacy.

(76) Absolute zero is of great significance to thermodynamics.

(77) That outcome is consistent with the prediction derived from the characteristics of the π-acceptor.

(78) The nitrile functionality is the synthetic equivalent of the carboxylic acid functionality.

(79) Ring opening polymerization accompanied by C-C bond scission produces linear polymer.

(80) The reactivity of the 1,3-dipole was ascertained by these experiments.

Lesson 9

(81) Absolute asymmetric syntheses are reactions using asymmetric sources such as circular polarization of light and a magnetic field, which are chiral but not chemical substrates.

(82) The results are consistent with the formation of a dimeric complex.
Point 「〜と矛盾しない，一致する」は be consistent with 〜．
この表現が定番．
Jeff's advice

(83) They have disclosed novel methodology to suppress the emission of air-polluting substances.
Point method は可算名詞である．一方，methodology を「方法論」という意味で使うときは抽象名詞(不可算名詞)．

(84) Photochromism can be defined as a reversible change of two chemical species with different absorption spectra induced by electromagnetic radiation.

(85) The electrolysis of water generates dihydrogen at the cathode.

(86) Intramolecular $\pi-\pi$-stacking plays a crucial role in the stabilization of the macromolecule.

(87) A company has demonstrated that the nematic polymer can be employed as a material for liquid crystal displays.

(88) That paper refers to a fact that is unbelievable but reliable.

(89) His perception of the danger was unsatisfactory.

(90) This syringe is disposable.

Lesson 10

(91) She fulfilled all of the requirements for getting a Ph.D.

(92) Ether is a flammable and harmful liquid.
　Point 不燃性は nonflammable あるいは incombustible．

(93) They confronted the issue of recycling reagent bottles.

(94) The paramagnetic nature rules out the possibility of the structure.

(95) The reagent diffuses into the porous zeolite.

(96) When a lens is thicker at the edges than it is in the middle, it tends to make light rays divergent.

(97) Neither of the two assumptions is correct.
　Point 二つのものについての否定で neither を使うときは「単数」扱い．

(98) Each atomic orbital is quantized and its energy is intrinsic to each element.

(99) The differential coefficients of the energies both at the transition state and the intermediate are zero.

(100) Radicals are chemical species with SOMOs.

Chapter 7

Lesson 1

(a) ナノテクノロジーは，大まかにいって，通常1から100ナノメートルの範囲のナノスケールの規模をもつ物質や構造物を取り扱うことを意味する．おそらく無意識のうちにわれわれは，すでに日常生活のなかでいくつかナノテクノロジーらしきものに遭遇しているであろう．

(b) ナノテクノロジーは単に非常に小さい物のサイズに関するものではない．より重要なことは，ナノテクノロジーは原子あるいは分子のレベルで作用する構造およびその能力に関するものであるということだ．つまり，それらのサイズと構造のために，新規で非常に特異な物理的，化学的，そして生化学的特徴をしばしば示す物質群や系を生じるのである．

(c) これらのナノ構造は，どのようにしてつくられたか，またどのように原子や分子が配列しているかという点で異なる．たとえばナノ粒子は，——およそ1～100 nmの直径の何万もの原子の集まりであるが——ナノテクノロジーにおける最も基礎的な構造である．

(d) ナノ構造はしばしば，水素結合，双極子力，親水性あるいは疎水性相互作用，重力そして他の力に起因して，配向したり，高度に配列した層に自己集合したりする．膜，ベシクル，DNAのような多くの天然にみられる生物学的な構造は，そのような自己集合化により生じる．

Lesson 2

　私は1988年の8月から2年間，ドイツの名門校，ハイデルベルグ大学のハインリッヒ・ウィーランド教授の研究室で研究生活を過ごした．これには妻と子供も同伴した．研究だけでなく，ドイツ語を学んだり，ドイツの人々や文化，慣例にふれたりと公私とも充実した楽しい生活であった．研究では，ルテニウムポリピリジン骨格をもつ多核錯体を合成することに成功した．大学の研究設備は整備されており，二人の優秀な博士課程の大学院学生にも恵まれた．互いに，技量と知識を交換しあえたものと思う．この留学で得た経験は，自国の大学での講義，大学院学生に対する指導，実験データの解析などの今後の教育と研究活動におおいに役立つであろう．(298字)

Lesson 3

(1) おそらく化学において新たな叡智が発展するときの特徴とは，ほかのどのような科学分野にも増して，一つには間違いなく危険のない予見しうる理論と，他方では危険でもあり驚くべき現実との間の目をみはるような相互連携である．

　Point the development of ～ has been characterized ～ は直訳すると訳しにくいので，「～の特徴は～である」と，日本語らしい構文になるよう意訳するとよい．

(2) (i) 有益なことを偶然発見し実用すること．(18 文字)

(ii) 1967 年にアセチレンの重合について研究していた日本の研究グループが，予定よりはるかに大量の触媒を用いて重合反応を行った．その結果，光沢のある銀色のフィルムを得た．その後，それをヨウ素で酸化することでいろいろな用途に使用可能な伝導性プラスチックが得られることを見いだした．この発見を実用するには物理学者と化学者が互いに協力することが必要であった．(173 文字)

Lesson 4

(1) (a) 太陽光の強度の高い紫外線が水を分解し水素とともに酸素を生成した．これにより酸素濃度レベルは現在の 0.1% にまで達したが，酸素自身が紫外線を吸収したため，この反応はここで停止した．(87 文字)

(b) 光合成をすることのできるバクテリアが可視光を利用して，二酸化炭素を炭水化物と酸素に変換した．これにより現在の 1% の酸素濃度レベルにまで達した．(71 文字)

(2) 酸素を取り込むこと．(10 文字)

(3) (i) そのレベルが今日の値の 10% にまで達したとき，上空にオゾン層が十分に発達し，その結果，地球の表面をより殺傷力の強い紫外線(C 波紫外線)から保護した．この重要な段階は約 4 億年前に達成され，このときから生命は地上で進化することが可能になった．

(ii) 植物体の発展は，光合成が進行するにつれて二酸化炭素の割合を減少させ，酸素の割合を増加させたので，大気の主要成分は酸素と窒素となり，二酸化炭素はごく少量となった．

Lesson 5

(a) The nature of gas is characterized by its pressure and temperature.

(b) However, the average velocity of a number of molecules can be considered to be unchanged under an equilibrium state.

(c) That is, we suppose that when molecules collide with each other, the energy of translational movement is not converted into electronic, vibrational, or rotational energy.

(d) The volume of molecules in a container is negligible compared to the volume of the container under a dilute atmosphere.

Lesson 6

(1) The first ionization energy of an atom is the minimum energy required to convert one mol of gaseous atoms in the ground state into one mol of gaseous monovalent cations. On the other hand, the electron affinity is defined as the internal energy change caused by the addition of electrons to one mol of gaseous atoms to produce one mol of gaseous anions.

(2) Carbocations react with electron rich compounds such as alkenes, enol ethers, enamines, and aromatic compounds, to form carbon-carbon bonds. Synthetically versatile carbocations can be categorized into three classes on the basis of their oxidation levels. The first group is alkylcarbocations, which can be generated by heterolysis of alkyl halides or alcohols. The second group is derived from aldehydes or ketones. The last group is acylium ions produced by bond scission of acid derivatives in the presence of a Lewis acid catalyst.

> **Point** fission は「核分裂，細胞分裂」におもに用いる．scission は「化学結合の分裂」，cleavage はより一般的に「分裂」を表すときに用いる．

Lesson 7

(1) 科学革命は，「新たな道具を用いることによってなされる革命」と「新たな概念を創出することによってなされる革命」の二つの種類に大別できる．後者は，大衆の大きな関心を集め，インパクトの強いものであるが，その数は前者に比較すると少ない．前者は，一般大衆にはそれほど大きな印象を与えるものではないが，科学の進歩にとっては同じく重要であり，多くの新発見を含んでいる．

(2) 著者らは，天文学や生物学などの分野で「新しい道具を用いること」によってなされる科学革命は，「新概念創出」による科学革命と同じ程度，あるいはそれ以上に重要であるとしているのに対し，Thomas Kuhn は量子力学に代表される，「新概念創出」による科学革命をほかにも増して重要であるとしている．

(3) (a) Kuhn の書いた本は非常に素晴らしく書かれていたので，すぐにその時代の名著となった．

(b) 物理学のなかで発見されたことを理解するための新しい概念を創出することに，最近の物理学は，あまり成功していない．

参考図書

1) 今村　昌著，『化学系の英語入門——読む・書く・話す英語の基礎』，(講談社，1955 年)
2) 福馬淳子著，『話しながら学ぶ化学英語』，(廣川書店，1986 年)
3) 足立吟也ほか著，『化学英語の活用辞典(第 2 版)』，(化学同人，1999 年)
4) 岩田　薫ほか編，『科学・技術英語例解辞典』，(三共出版，2003 年)
5) 亀井エリザベス著，『なっとくする基礎科学英語』，(講談社，1999 年)
6) 日本化学会化合物命名小委員会編，『化合物命名法(補訂 7 版)』，(日本化学会，2000 年).
7) 小川雅彌・村井眞二監修，『有機化合物命名のてびき』，(化学同人，1990 年)
8) 『SENTENCE CD-ROM 版』，(朝日出版社，2002 年)
9) 大澤善次郎著，『化学英語の手引き』，(裳華房，1999 年)
10) 鷲見由理著，『CD 付き 英語の発音が正しくなる本』，(ナツメ社，2007 年)
11) 松澤喜好著，『英語耳——発音ができるとリスニングができる』，(アスキー，2004 年)
12) 巽　一朗著，『英語の発音がよくなる本』，(中経出版，2005 年)
13) 原田豊太郎著，『間違いだらけの英語科学論文』，(講談社，2004 年)
14) 鈴木英次著，『科学英語のセンスを磨く——オリジナルペーパーに見られる表現』，(化学同人，1999 年)
15) 大井静雄著，『国際学会英語表現辞典——Congress English』，(三輪書店，1998 年)
16) 平田光男著，『科学英語の基礎——これならわかる英文読解術』，(化学同人，2003 年)
17) ロバート・M. ルイスほか著，『科学者・技術者のための英語論文の書き方——国際的に通用する論文を書く秘訣』，(東京化学同人，2004 年)
18) 野口ジュディー著，『Judy 先生の耳から学ぶ科学英語』，(講談社，1995 年)
19) 友清理士著，『理化学英語の冠詞の用法』，(研究社，2004 年)
20) 原田豊太郎著，『例文詳解 技術英語の冠詞活用入門』，(日刊工業新聞社，2000 年)

INDEX

■ A ■

ability 110, 112
absence 89
   in the absence of ～ 47
absolute 38, 97, 98
   absolute asymmetric synthesis 163
   absolute configuration 96, 122
   absolute temperature 56
   absolute zero 163
absorption 163
abstract 48
academic year 35
accelerate 60
acceptance 121
acceptor 158, 159
accident 71
accompany 163
accord
   be in accord with ～ 130
accordance
   be in accordance with ～ 77
according
   according to ～ 71, 161
account
   account for ～ 88
accurate 53
acetal 27
acetaldehyde 26
acetic acid 27
acetone 26, 121
acetonitrile 27
acetylene 25
acetylsalicylic acid 29
acetyl chloride 29
achieve 121, 129, 157, 161
achievement 73
acid 162
acrolein 29
acrophobic 21
acrylonitrile 27
ACS 121

act
   act as ～ 115
actinium 23
activate 157
activation 62, 103
   activation energy 39, 58, 160
active 51
actively 59
activity 63
acylation 118
addition 43, 60, 61, 86, 94, 95, 97, 101, 159
   in addition 121
additional 89, 118
additive 50
adipic acid 29
adjacent 121
   be adjacent to ～ 59
adjust 120
advanced 81
advent
   with the advent of ～ 73
affect 56, 161
afford 13, 127, 128, 131
agostic interaction 34
air 4
air-polluting 163
alanine 30
alcohol 26, 47, 88, 118
aldehyde 26, 47, 93, 95
aldol reaction 31
aliphatic 159
alkane 3, 7, 24, 88
alkene 25, 85, 114, 115
alkyl
   alkyl group 160, 162
   alkyl halide 100, 101
alkylation 118
alkyne 7, 25
allow 40, 127
alloy 119, 162

allyl 16
   allyl bromide 29
already 50, 70
alternative 121
aluminum 5, 12, 22
alumni association 35
ambient temperature 40
ambiguous 74
amide 93
amine 27
amino acid 30, 112
ammonia 30, 43, 160
amount 162
amphiphilic 20, 68
amphoteric 20
analysis 53, 74, 124
analytical chemistry 35
analyze 53, 55, 157, 159
anhydride 93
anhydrous 48
aniline 29
anion 48
anisole 29
annually 108
anthracene 28
antibiotic 159
antibody 162
anticipate 53
anticlockwise 98
antimony 23
apart from ～ 82
apparatus 52
appear 86
application 121
applied chemistry 35
apply 75, 99, 118
   apply ～ to ～ 72
approach 107, 121
appropriate 121
aqueous 48, 110
   aqueous solution 43
architect 106

architecture   106
argon   22
arise   97
aromatic ring   58
Arrhenius equation   31
arsenic   23
artificially   118
aryl   16
   aryl iodide   29
ascertain   77, 163
aspect   77
aspirator   36
aspirin   29
assembly   158
assign   97, 98
assignable
   be assignable to ∼   53
assignment   96, 98
associate
   be associated with ∼   81, 112
assumption   72, 164
astatine   23
asymmetric   51, 129
   asymmetric amplification   34, 161
   asymmetric compound   161
   asymmetric molecule   161
   asymmetric multiplication   34, 161
   asymmetric source   163
   asymmetric synthesis   162
as for ∼   50
atmosphere   87, 108
atom   81, 82, 83, 84, 85, 86, 88, 95, 96, 158, 159, 160
atomic   84, 86
attach   95, 98
attain   129
attempt   6, 39, 52
attention   72
attract   88
attraction   88
attractive   62
attributable
   be attributable to ∼   71, 118
authentic   44, 162
auxiliary   51
available   40
aversion   121
avoid   48, 97

■ B ■

β-elimination   33
β-particle   84
bachelor's degree   35
Bachelor of Arts   35
back donation   33
bacteria   87
Baeyer-Villiger oxidation   31
balance   36
Baldwin rule   31
balm   3
ban   87
barium   23
barrier   103
base   48, 100, 114
   be based upon ∼   91
basic   105, 107
basis   110
   on the basis of ∼   53
beaker   36
bear   67
Beckmann rearrangement   31
belong
   belong to ∼   75
bench   36
benzaldehyde   26
benzene   28
benzoic acid   27
benzonitrile   27
benzophenone   26
benzyl alcohol   26
beryllium   22
binding energy   47
bind to   114
biochemistry   35
biodegradable   21
biological   91, 94, 107, 108, 110
biology   35, 108
bipyridine   20
Birch reduction   31
bis   20
bismuth   23
bit   5
bite angle   34
bleach   11
blood   68
boil   7
boiling point(b.p.)   58, 72, 88, 160
bond   15, 85, 86, 94, 99, 106, 121, 122, 163
   bond energy   85, 86
   bond order   58
bonding   99, 114, 115
boron   22
boss   161
bound   104
branch   159
brass   64
break   85, 86, 121
bridge   162
bridging complex   130
bring
   bring about ∼   55, 89, 162
bromine   23, 51
Brønsted acid   31
Brookhart catalysts   31
building block   106, 110
bulky   130
buret・burette   36
butadiene   58
butanal   26
butane   15, 24, 58
butanol   26
butene   25
butylaldehyde   26
butylene   25
butyne   25

■ C ■

γ-radiation   84
cadmium   23
Cahn-Ingold-Prelog(CIP)   96
calcium   22
calm   3
candidate   62
carbocation   158
carbohydrate   108
carbon   8, 22, 83, 84, 86, 95, 96, 120
   carbon dioxide   108
   carbon monoxide   19
carbonyl   93, 94, 95

carbonyl compound 93
carbonyl group 51, 59, 94, 97, 98, 101, 121
carbopalladation 34
carboxylic acid 27, 93, 163
carry
　carry out 43, 51, 118, 127, 160
catalysis 89
catalyst 39, 43, 47, 62, 63, 89, 91, 92, 118, 121, 161, 162
catalytic 118, 158, 159, 161
　catalytic constant 91
catalyze 61, 89, 90, 101, 110, 130
categorize 70
category 75
cathode 163
cause 6, 57
cell 110, 112, 162
cellular 110
cellulose 67
center 121
centrifugation 157
cerium 119
certain 103
cesium 23
chain 105, 159
　chain reaction 160
challenging 161
change 52, 55, 163
chapter 110
characteristic 112, 163
charge 11, 83, 158
charged 81, 82, 83
chemical 50, 68, 73, 75, 106, 110, 118, 163, 164
　chemical engineering 35
　chemical equation 89
chemist 77, 81, 82, 106, 161
chemistry 85, 86, 94, 110, 159
chemoselective 160
chiral 51, 95, 118, 122, 163
　chiral center 96, 121
　chiral ligand 162
chirality 96

chlorine 22
chlorofluorocarbon 87
chloromethylpropane 102
chloroprene 29
chromatography 36, 53
chromium 22
circular polarization of light 163
claim 77
Claisen condensation 31
Claisen rearrangement 31
clarify 66, 118
class reunion 35
clearly 130, 157
Clemmensen reduction 31
cloud 158
cluster 62
co-catalyst 62
coal 67
cobalt 22
coefficient 164
collate 75
collision 103
combination 55
combinatorial 121
combine 106
commencement 35
commercially 40
commonly 91
community 121
comparable
　be comparable to ～ 68, 160
comparative 86
compare 88
competence 120
complete 71
completion 62
complex 40, 54, 61, 93, 110, 121, 124, 163
component 45, 93, 106, 107, 119, 158
composition 159
compound 3, 6, 39, 41, 48, 53, 55, 61, 71, 73, 87, 89, 91, 94, 110, 120, 121, 124, 125, 126, 129, 158, 159, 160, 161, 162
comprise 158
　be comprised of ～ 67
compulsory education 35

conceivable 71
concentration 59, 92, 102, 119
concept 76, 112, 115, 120
concern
　be concerned with ～ 58, 95
conclude 157
conclusion 10, 77
conclusive 70
concomitant 112
condensation 94, 112
condenser 36
condition 41, 48, 49, 91, 102, 124
conduct 161
conductive 65, 157
conductivity 60, 68
cone angle 34
configuration 97, 98, 124
configurationally 122
confirm 53, 161
conflict
　be in conflict with ～ 77
conformation 62, 160
confront 164
confusion 97
conjugate 93
　conjugate ～ with ～ 46
conjugation 60
connection 85
consider 77, 102
　consider ～ as ～ 48
considerable 62
consist
　consist of ～ 64, 68, 81, 108, 162
consistent
　be consistent with ～ 130, 163
constant 56, 90, 92
construct 105, 112
consume 59, 159
consumption 108
contain 12, 45, 82, 83, 84, 88, 162
context 93
contraction 112
contrast
　in contrast 94
　in stark contrast 39

contribution 73, 114
control 57, 118, 160
  out of control 60
convention 96
conventional 76
convert 61, 160
convince 77
cool 162
coordination 101
Cope rearrangement 31
coplanar 58
copper 22, 64
correct 164
corresponding 58, 100
correspond to ∼ 112
costly 119
coupling constant 75, 160
covalently 104, 106
covalent bond 106
Cram's rule 31
create 106, 121
credit 34
cresol 29
crucial 130
crude 72, 74, 157
crystal 73, 121, 122, 158
crystalline 121
crystallize
  crystallize as ∼ 74
crystallographic 124
crystallography 73

■ D ■

damage 87
danger 164
Darzens reaction 31
data 75, 77, 130, 157, 160
dean 35
deca 19
decane 24
decarbonylation 121
decision 10
decompose 48, 84
decomposition 39, 162
define 58, 82, 85, 106, 163
definition 162
degradation 75
degrade 110

degree 104
dehydrate 110
dehydration 110
delocalization 47
demonstrate 68, 164
denote 83, 84
density 114
depend
  depend on ∼ 158
dependence 118
dependent
  dependent on 108
derivative 93, 110, 118, 121
derive
  derive ∼ from ∼ 72, 104,
    108, 112, 163
describe 89, 107, 110, 121, 158
descriptor 97
descriptors 96
desiccator 36
design 106, 119
desire 121
detail 92, 96
determine 73, 92, 124
deuterated chloroform 29
deuterium 22
deuteron 22
develop 118, 160
device 14, 70
Dewar-Chatt-Duncanson model
  115
Dewar-Chatt bonding model 31
di 19
dialkylmagnesium 100
dichloromethane 29
Dieckmann condensation 31
Diels-Alder reaction 31, 118
diene 118, 159
differential 164
diffuse
  diffuse into ∼ 164
dihydrogen 163
dimensional 158
dimer 19, 105
dimeric 163
dimethyl malonate 27
dimethyl sulfoxide 41
dinitrogen 161

dioxane 28
dip 11
dipole 41, 88, 163
  dipole moment 19
disadvantage 20
disagree 20
discipline 120
disclose 70, 163
discovery 118
discuss 158
discussion 112
disease 162
disk 70
display 70, 164
disposable 164
dissolve 58
distill 45
distinction 94
distribution 66
divergent 164
DNA 70
doctor 4
doctorate 35
dodeca 19
dominate 85
donate 114
donation 114
dope 68
double bond 85, 86
dramatically 60, 121
drastic 45
drawback 69
drawing 97
drug 4, 68
dry 11
  dry over 38
due
  be due to ∼ 53
  due to 64, 162
dysprosium 119

■ E ■

e.e. 161
earth 108, 119
economical 119
effect 83, 86, 162
  effect on ∼ 101
effective 114, 118

effectively 91
effectiveness 91
efficacy 163
efficiency 45, 68
efficient 62
Einstein 72
elasticity 161
elective subject 35
electric 119
electrical 112
electrically 65, 82, 83, 161, 162
electrolysis 21, 163
electromagnetic radiation 163
electromagnetic wave 158
electron 47, 51, 81, 82, 83, 88, 114, 122, 158, 159
  electron-donating 46
  electron-withdrawing 46
  electron-withdrawing group 162
  electron acceptor 115
  electron donor 115
  electron pair 88, 157
electronegative 88
electronic 162
electrophile 101
element 83, 119, 162
elementary 61
elimination 86
elusive 51
emission 163
emit 160
employ 129, 161, 164
  employ ～ for ～ 69
employment 162
enantiomer 61, 95, 96, 97, 98
enantiomeric 121
enantioselectivity 118
encompass 120
encounter 45, 105
endothermic 157
energetic 103
energy 6, 98, 103, 108, 110, 112, 159, 160
engineering 13
enhance 60
enhancement 89
enthalpy 85

entity 81
entrance ceremony 34
entropy 6, 160
environment 87, 122
envision 45
enzyme 74, 91, 110
epoxide 28, 101
equal 55, 58, 86, 97, 98, 158
equation(eqn) 56, 91
equilibrium 56, 100
equivalent 58, 64, 108, 163
erbium 119
Eschenmoser reaction 31
ESR 162
essence 108
essential 62, 94
establish 50
ester 27, 53, 93
esterification 46
estimate 108, 159
ethanal 26
ethane 24, 88, 104
ethanenitrile 27
ethanoic acid 27
ethanol 26, 55
ethanolamine 118
ethene 25
ether 27, 38, 88, 100, 160, 164
ethylene 25, 64, 104
  ethylene glycol 26
ethyl acetate 27, 48
ethyl group 48
ethyne 25
evaporate 157
evaporator 36
evidence 87, 158, 159
evolution 108, 118
evolve 45
exact 124
examination 72
examine 118, 158
example 92, 103, 112
excess 121
exchange reaction 64
excite 160
exclude 75
exemplify 76
exert 62

exist 119
existence 118
exothermic 43, 130
expect 14, 86
expel 50
experiment 160, 163
experimental 77, 91, 130, 157, 160
experimentally 103
explain 122, 162
exploit
  exploit ～ for ～ 65
explore 70
expression 162
extend 60
extension 104
extent 45
extra 47
extract 45, 119
extraction 119
Eyring equation 32

■ F ■

face 97
facile 64
facilitate 110
factor 85
fail
  fail to ～ 73
favorable 112
feature 62, 106
featureless 81
Felkin-Ahn model 32
fiber 65
field 106, 120
fill 5
  fill ～ with ～ 43
film 66, 68
filter
  filter off 43
filtration 51
fingerprint region 161
first order 103
first order kinetics 102
fix 56, 108
flammable 164
flash 10
flask 43

Erlenmeyer flask   36
pear shape flask   36
two-necked flask   36
volumetric flask   36
fluorene   28
fluorine   22, 88
focus   112, 118
   focus on ～   65
follow   99
following   56, 70, 99, 102, 161
force   106
form   6, 97, 98, 99, 100, 104, 112, 121, 122, 128, 129
formaldehyde   26
formation   95, 101, 108, 121, 128, 129, 163
formic acid   39
fortunate   119
fraction   72
framework   95
francium   23
freezing point   58
frequency   158
freshman   35
Friedel-Crafts alkylation   32
fulfill   164
fullerene   28
fume hood   36
function   70, 102
functional   106
   functional group   51, 67, 93, 94, 118
functionality   163
fundamental   76, 100
funnel   36
   separatory funnel   36
furan   28, 159
furnace   162
furnish   4, 44, 49, 128

■ G ■

gallium   23
gap   160
gas   3, 45, 53, 56, 158
gasoline   72
Gaussian distribution   66
GC-MASS   159
general   121

   in general   46, 122
generalize   107
generate   110, 128, 160, 161, 163
generic   119
genetic   70, 112, 118
germanium   23
give   92, 100, 105, 128
glue   106
glycerin   26
glycerol   26
go
   go off   99
gold   8, 23, 119, 159
govern   100
grad   35
grade   3, 34
gradient   92
graduate   35
   graduate school   35
graduation ceremony   34
graduation thesis   35
graph   92
grasp   107
Greek   104
Grignard reagent   32, 47, 95, 100, 101
ground state   160
group   5, 41, 46, 95
Grubbs catalyst   32

■ H ■

hacker   14
hafnium   23
half-life   84
halide   3
halogen   86
Hammet relationship   32
Hammond postulate   32
hapticity   33
harmful   164
harness   110
harsh   49
Hartwig-Buchwald coupling   32
heat   5, 14, 39, 45
Heck reaction   32
helium   22, 82
hepta   19

heptakis   20
heptane   24
heptene   25
heptyne   25
heterolysis   158
hexa   19
hexakis   20
hexane   24
hexanone   26
hexene   25
hexyne   25
histamine   68
hold   14
   hold together   106
holmium   119
homogeneous   121
honor   118
Horner-Wadsworth-Emmons reaction   32
hot   3
Hückel's rule   32
hydroboration   34
hydrochloric acid   30
hydrogen   22, 82, 160
   hydrogen bond   88
   hydrogen bonding (H-bonding)   49, 88, 118, 160
   hydrogen peroxide   30
hydrolysis   21, 102
hydrometallation   34
hydrophilic   21, 68
hydrophobic   21, 68
hyperconjugation   162
hypothesis   72, 118

■ I ■

ideal   56
identical   98
   be identical to ～   76
identify   54, 162
identity   82
if anything   58
illustrate   89
imidazole   28
immediately   160
impact   103
implication   72
impossible   157

improve 5, 45, 120
inactivate 161
include 94, 112
inclusion 20
incorporation 161
increase 121
independent
　be independent of 〜 75
　independent administrative
　　corporation 34
indicate 39, 91, 118, 159
indication 107
indium 23
individual 106
indivisible 82
indole 28
induce 163
inductive effect 160
industry 73, 121
infer 160
influence 62, 161
infrared radiation 158
ingredient 119
initial 59
initially 118
injection 20
innovation 162
innovative 70
inorganic chemistry 35, 107
insist 55
insoluble 157, 158
inspiration 107
instead 103
institute 75
instruct 161
intact 45
interaction 62, 88, 118, 157
intercept 92
interconversion 110
interfere
　interfere with 〜 63
intermediate 20, 164
intermolecular 20, 106
intracellular 21
intramolecular 21, 163
introduce 110, 130
introduction 107, 130
invariably 91

inversion 160
investigate 65, 106
involve 85, 88, 91, 95, 106,
　110, 121
in other words 41, 105
in situ 52, 161
in vitro 118
in vivo 68
iodine 23, 54
IR 161
iridium 23
iron 22
irreproducibility 45
isolate 129
isoleucine 30
isomer 45, 162
isoprene 28
isotope 83, 84
issue 164

■ J ■
jack 11
Jahn-Teller distortion 32
jargon 11
Jones oxidation 32
judge
　judge from 〜 163
junior 35

■ K ■
Kaminsky catalyst 32
Karplus equation 32
Kelvin 56
Kepler's law 72
ketone 8, 26, 48, 93, 121
kinetic 57, 59, 61, 118
kinetics 92, 102
knowledge
　to (the best of) our knowledge
　　75
Kolbe coupling 32
krypton 23

■ L ■
laboratory 35
lanthanum 23
large 4
laser crystal 119

law 89
layer 45
lead 23, 55
leaving group 50, 94, 99
lecturer 35
LED 124
leucine 30
Lewis acid 32, 101, 115
Lewis base 115
liberal art(s) 35, 120
ligand 62
　ligand field theory 33
lignin 67
Lindlar catalyst 32
line
　line up 82
linear 163
link 104, 118, 160
lipophilic 21
liposome 21
liquefaction 158
liquefy 21, 158
liquid 43, 164
　liquid crystal 70, 164
listening comprehension test 34
lithium 22
　lithium hydride 30
living creature 84
locate 35
loss 112
lower 58

■ M ■
machine 52
　machine scored exam 34
macromolecule 104, 163
magnesium 22, 100, 101
　magnesium dibromide 100
　magnesium oxide 30
magnetic field 75, 163
magnetic stirrer 36
magnification 120
magnitude 83
maintain 48, 75, 121
major 159
　major in 35
make
　make up 104

maleic anhydride 29
manganese 22
　manganese dioxide 30
mankind 108
Mannich reaction 32
manufacture 119
Markovnikov rule 32
mass 81, 83, 84, 108
　mass spectrometry 161
master's degree 35
mastery 120
material 69, 70, 121, 164
　material chemistry 35
McMurry coupling 32
mean 104, 105
measure 10, 45, 56, 74, 91, 92, 124, 158
measurement 162
mechanical 112
mechanism 12, 40, 43, 50, 51, 75, 99, 102, 118, 158
mediate 121
medical science 35
medicine 163
Meerwein-Ponndorf reduction 32
menace 87
mercury 23, 119
Merrifield method 32
meta-orientation 162
metabolism 110
metabolite 110
metal 62, 114, 115, 119, 162
　metal ion 157
metallacycle 34
metallic 38, 119
metathesis 34
methacrolein 29
methacrylic acid 29
methanal 26
methane 24
methanol 26, 88
methionine 31
method 53, 54, 77, 92, 161, 163
　method for ～ 43, 48, 160
methodology 70, 76, 161, 163
methoxy group 46, 160

methyl 9
　methyl bromide 29
methylene chloride 29
Michael addition 32
micromolecule 104
microscope 36, 120
millimeter 82
mineral 119
Mitsunobu reaction 32
mix 162
mixture 11, 40, 43, 48, 74, 97, 125, 157, 158
modern science 72
modify 93
moiety 48, 68
moisture 100
molar mass 105
mole 56
molecular 66, 95, 106
　molecular capsule 162
　molecular mass 104
molecule 12, 42, 68, 94, 103, 104, 106, 112, 121, 158
molybdenum 23
moment 41
mono 19
monolayer 19
monomer 105, 112
Monsanto process 32
motif 121
motor 119
multidisciplinary 106
muscle 112

■ N ■

naphthalene 28
national center test for university administration 34
naturally 84
natural science 35
nature 110, 112, 158, 164
needle 11, 36, 74
negatively charged 81, 83, 158
negligibly 58
nematic 164
neodymium 119
neon 22
nerve 112

neutral 82, 83
neutron 82, 83, 84
Newman projection 32
Nichrome™ 68
nickel 22
niobium 23
nitric acid 30
nitrile 27, 163
nitrobenzene 46
nitrogen 22, 45, 88
nitromethane 29
nitro group 46
NMR 64, 74, 157, 158
Nobel Prize 118
non-bonding 157
nona 19
nonane 24
nonbonded 88, 99
nonene 25
nonyne 25
normal 92
novel 64, 118, 160, 163
no longer 49, 73
nucleic acid 112
nucleophile 50, 94, 99, 100
nucleophilic 21, 95, 97, 99, 101
　nucleophilic addition 47
　nucleophilic attack 157
nucleus 81, 82, 83, 84, 86, 158
nylons 65

■ O ■

obey 56, 102
object 3, 6
objective 120
observation 52
observe 103
obtain 42, 73, 128
obviously 130
occasional 103
occupy 56, 114
occur 50, 84, 89, 158, 159
octa 19
octakis 20
octane 24
octene 25
octyne 25
offer 121

olefin 114, 115
oligomer 21, 105
oligonucleotide 21
on the other hand 161
operate 52, 102
opinion 15
Oppenauer oxidation 32
optical 70, 161
optically 51
oral exam 34
orbital 114, 157
order 6, 82, 92
organic 3, 45, 121
   organic chemistry 35, 96, 107, 120, 121
   organic compound 110
   organic material 108
   organic reaction 100
   organic synthesis 93, 94, 108, 121
organolithium 100
osmium 23
outcome 7, 57, 163
overall 94, 108
overcome 103
overlap 114
owing to ~ 77
oxidative 61
   oxidative addition 33
oxide 119
oxygen 22, 56, 63, 86, 88, 100, 157, 160
ozone 30
   ozone layer 87
ozonolysis 158

■ P ■

π–allyl 34
packing coefficient 158
palladium 23
palm 3
paper 161, 162, 164
paramagnetic 164
partial 56
participate 118, 162
particle 82, 158, 159
particular 72, 91
   in particular 118

patent 70
pathway 89, 90
Pauli exclusion principle 32
Pauson-Khand reaction 32
peak 5, 53
penta 19
pentakis 20
pentane 24
pentene 25
pentyne 25
percent 4
perception 164
perform 118, 127, 161
periodic table 162
Perkin reaction 32
permanent 5
   permanent magnet 119
permit 52
perpendicular 58
Peterson elimination 32
petri dish 36
petroleum 67, 72
Ph.D. 164
pharmacy 35
phenol 26, 159
phenomenon 161
phenylalanine 31
phosphate 110
phosphine 64
phosphoric acid 30
phosphorus 22
photochemical 108, 121, 157, 160
photochromism 163
photocrosslinkable 65
photoirradiation 161
photosynthesis 108, 157
physical 106
   physical chemistry 35, 107
physics 35
picric acid 29
pipet 36
plan 71
plane 97, 98
planet 72, 108
plant 108
platinum 23
play

play a ~ role 110
play a ~ role in ~ 72, 163
pm 82
polar 4, 41
polonium 23
polyacetylene 68
polyester 65
polyethylene 104
polymer 65, 66, 67, 104, 105, 112, 121, 161, 163, 164
   polymer chain 104
polymerization 104
polypyrrole 66
polysaccharide 112
poor 101
porous 162, 164
position 56
positive 162
positively charged 81, 82, 158
possess 81, 82, 83, 84
possibility 164
possible 14, 49, 51, 118, 119, 159
postdoctoral fellow 35
postgraduate 35
potassium 22
   potassium cation 161
   potassium cyanide 30
potential 94
pour 48
practical 43, 75
precipitate 43
precise 74
precisely 83, 112, 114
prediction 163
predominantly 162
preparation 43, 48, 65, 66
prepare 121, 162
presence 89, 91, 101, 162
president 35
pressure 45, 56, 158
prevent 75
price 69
primary 110
principal 93
principle 107
prior
   prior to ~ 50

priority 98
procedural 92
procedure 43, 44, 76, 91, 97
proceed 47, 127
process 64, 84, 85, 103, 108, 112
produce 4, 45, 97, 108, 125, 126, 128, 158, 161, 162, 163
product 54, 61, 74, 95, 118, 119, 121, 159, 161
　product of 〜 and 〜 56
production 64
professor 35, 121
　assistant professor 35
　associate professor 35
profound 101
progress 72
　in progress 59
prominent 72
promising 70
proof 70
propane 24
propanenitrile 27
propanoic acid 27
propene 25
property 93, 106, 107, 118
proportion 84
proportional
　be proportional to 〜 56
proposal 158
propose 64, 75, 115
propylene 25
propyne 25
protein 73, 112
protium 22, 58
proton 22, 48, 82, 83, 158
prove 115
provide 39, 43, 70, 89, 107, 110, 112, 118, 128
pseudohalogen 21
pseudorotation 21
purity 53
purpose 66, 69
pursue 52
pyridine 28
pyrimidine 28
pyrrole 28

■ Q ■
quantitatively 42
quantity 97
quantum mechanics 35
quartz 162
quaternary 121
quench 48
quinoline 28

■ R ■
racemic 97, 118
radical 121, 122, 158, 160, 162, 164
radium 23
radius 81, 82
radon 23
raise 58
range
　a range of 〜 91, 107
ranging
　ranging from 〜 to 〜 72
rapidly 51
rare 119
　rare earth 119
rate 3, 89, 90, 98
　rate-limiting step 39
　rate constant 91, 92
　rate law 90
ratio 91
ray 164
react 94, 103
　react with 〜 41, 100, 126, 127
reactant 103
reaction 10, 39, 40, 41, 42, 43, 45, 46, 47, 48, 49, 50, 51, 58, 59, 60, 61, 62, 72, 74, 75, 76, 86, 89, 90, 91, 92, 93, 94, 95, 98, 99, 100, 101, 118, 121, 124, 125, 126, 127, 129, 130, 131, 157, 158, 159, 160, 161, 162, 163
　reaction mechanism 159
　reaction rate 42, 161
reactivity 100, 101, 163
reagent 7, 40, 45, 50, 100, 164
　reagent bottle 36, 164
realizable 157

receptor 161
recognize 161
recombination 160
record 6, 102
recrystallize 55
rectangle 58
recycling 21, 164
red-shift 55
reduce 8, 45, 50, 59, 108
reduction 108
reductive elimination 33
redundant 162
refer
　refer to 〜 38, 160, 164
　refer to 〜 as 〜 96
reflux 100
Reformatsky reaction 33
regard 81
　regard 〜 as 〜 48
regenerate 162
reject 77
relate
　be related to 〜 110, 112
relative 104, 105
relatively 101
release 84
reliable 21, 164
remain 42, 45, 51, 74, 89, 90, 162
remove 121, 157
repeat 104
replace 67
report 66, 159
represent 104, 108
reproducibility 163
require 59, 103, 107, 110, 119
required subject 35
requirement 164
research 52, 75, 159
researcher 75
residual 55
resolution 61, 118
resonance effect 160
respect
　with respect to 〜 115
respectively 49, 131
response 120
responsible 96

INDEX 179

result 4, 9, 39, 45, 72, 118, 121, 124, 128, 129, 159, 160, 163
   result in ~ 43, 47, 162
reuse 21
reverse 86
reversible 46, 160, 163
reversibly 106
re face 95, 98
rhenium 23
rhodium 23
ribonucleic acid (RNA) 118
rigorously 75
ring-opening metathesis polymerization 34
ring opening polymerization 163
RNA 118
Robinson cyclization 33
route 5, 121
rubber stopper 36
rubidium 23
rule 72
   rule out ~ 164
ruthenium 23

■ S ■

σ–allyl 34
safety glasses 36
salt 6
sample 44, 120, 129, 160, 162
   sample tubing 36
saturate 48
   saturate ~ with ~ 43
save 14
scale 36
scandium 22
scheme 45
Schlenk equilibrium 100
Schrock carbene 33
science 7, 107, 120
scission 163
scope 3
second order 103
section 99
selection 118
selectively 61
selenium 23

semester 34
senior 35
sensitive 54
separate 6, 52, 118, 158
sequence 98
serve 54
Sharpless asymmetric epoxidation 33
Shiff base 33
show 85
side 58
signal 112
significance 163
significantly 60, 119, 158, 161
silicon 22
silver 8, 23, 119
similar 70, 88, 97, 99, 129
Simmons-Smith reaction 33
singlet 13, 162
single bond 58
size 85, 86
si face 95, 98
slide 120
smoothly 47
soap 7, 68
sodium 7, 22, 38
   sodium bicarbonate 30, 48
   sodium chloride 30, 43
   sodium hydroxide 30
soil 7
solid 55, 157, 158
   solid-phase 121
   solid-state 121
solidify 21
solution 5, 45, 55, 101, 110, 121, 158, 160
solvent 58, 91, 103, 118, 157
solvolysis 159
SOMO 164
Sonogashira coupling 33
sophomore 35
source 103, 110
spatula 36
species 163, 164
spectral 52
spectrochemical series 34
spectrometer 74
spectrometry 159

spectroscopy 157
spectrum 53, 55, 64, 74, 158, 161, 162
spherical 81, 159
spin 58
split 108
spontaneously 159
square 58
stabilization 163
stabilize 114
stable 100
stacking 163
stage 11, 120
stand 69
standard 129
starch 54
starting material 48, 59, 159
steel cylinder 36
step 14, 39, 121
stereocenter 96
stereochemistry 160
stereoisomer 52
stilbene 28
Stille coupling 33
stimulate 118
stir 40, 125, 158
   stir bar 36
stoichiometric 158
storage 118
straightforward 121
strain energy 162
strength 85
stretching vibration 53
strictly 48
strontium 23
structural 121
structure 55, 73, 74, 81, 94, 97, 124, 158, 164
styrene 28
subject 11, 159, 161
   be subject to ~ 51
substance 58, 69, 161, 163
substantiate 77
substituent 95, 118, 121, 130
substitution 51, 94, 99, 101
substrate 118, 121, 162, 163
succeed 5
succinic acid 29

sugar   10, 110, 112
suggest   162
sulfoxide   51
sulfur   22, 162
sulfuric acid   30
sum   89
superconducting magnet   119
supramolecular   106, 107
   supramolecular chemistry   106
supply   103
suppress   163
surfactant   68
surround   81, 82, 158
susceptible   159
Suzuki coupling   33
Swern oxidation   33
syllabus   34
symmetry   114
synthesis   47, 51, 71, 107, 121, 161
synthesize   67, 106, 110, 161
synthetic   65, 121, 163
   synthetic chemistry   35
syringe   36, 164
system   53, 103, 107, 112, 160
systematic   159

■ T ■

tab   15
table   10, 85, 86
tablet   10
take
   take place   40, 98, 124, 127, 159, 162
Tamao coupling   33
tantalum   23
target   4, 42
Tebbe's reagent   33
technetium   23
technique   118
tellurium   23
temperature   56, 64, 69, 75, 91, 100, 119
term   4, 34, 38, 85, 104
   in terms of ～   58
terminate   160
terphenyl   20
tert-butyl cyclohexane

carboxylate   27
test   54
   test tube   36
   test tube rack   36
tetra   19
tetrahedral   95
tetrakis   20
thallium   23
theoretical   77
thermodynamic   57
thermodynamics   163
thermometer   36
thiazole   28
thick   164
thin   9
thionyl chloride   30
thiophene   28
three-way stopcock   36
tin   23
tissue   68
titanium   22
tolerate   51
toluene   28
torsional   159
transfer   51, 159
transformation   76, 77, 110
transition
   transition-metal   61
   transition metal   114, 121
   transition state   98, 160, 164
transmetalation   33
trap   122
treatment   45, 129, 162
tri   19
triangle   19
trigger   8, 51
trimer   105
triphenylphosphine   29
   triphenylphosphine sulfide   29
triplet   19, 162
tris   20
tritium   22
triton   22
tryptophan   31
tube   4
tuition   34
tungsten   23, 119
tweezers   36

type   85
tyrosine   31

■ U ■

ubiquitous   119
UK   87
ultrasonic cleaner   36
unambiguous   20
unambiguously   73
unbelievable   164
uncatalyze   89, 90, 91, 92
unchanged   42, 89
undeca   19
undergo   39, 94, 100, 101
undergraduate   35, 96
undetectable   91
unfavorable   112
unique   62, 161
unit   104, 105, 112
university
   national university   34
   private university   34
   state university   34
unpredictable   72
unsatisfactory   164
unsaturation   20, 93
unsuccessful   52
uranium   23
urea   27
utility   76
UV absorption spectrum   124
UV ray   158

■ V ■

vacuum pump   36
vague   8
validity   72
valine   30
value   77, 92, 104, 115, 121
vanadium   22
various   118
versatile   62
vessel   162
via   108
viable   121
vicinal   121
view   81, 104
vigorously   45

vinegar 14
vinyl chloride 29
virtually 119
virtue
　by virtue of ～ 98
visible light 158, 160
vision 10
volume 56

■ W ■

Wacker oxidation 33
warm 6
waste 15
water 108
wavelength 158
weigh 82
weighing paper 36

weight 66, 161
well
　as well as 65, 112, 158
widespread 95
Willkinson's catalyst 33
wire 68
Wittig reaction 33
Wolf-Kishner reduction 33
Wolf rearrangement 33
Woodward-Hoffmann rule 33
work
　work out 92
worth 72, 88
Wurtz reaction 33

■ X ■

X-ray 73, 124, 158

xenon 23
xylene 28

■ Y ■

yeast 15
yield 15, 49, 101, 121, 128, 131, 161
yttrium 23

■ Z ■

Zeise's salt 33
zeolite 162, 164
zero 164
Ziegler-Natta catalyst 33
zinc 9, 22, 64
zirconium 23
zoom 9

● 著 者

國安　均（くにやす　ひとし）
1964年大阪府生まれ．1993年大阪大学大学院工学研究科博士課程修了．2007年，大阪大学大学院工学研究科准教授．2021年逝去．専門は有機金属化学．

● 英語監修者

ジェフリー・M・ストライカー(Jeffrey M. Stryker)
1956年インディアナ州生まれ．1983年コロンビア大学(Columbia University)で博士号取得．現在，カナダのアルバータ大学(The University of Alberta)化学科教授．専門は有機金属化学．

化学英語101　　リスニングとスピーキングで効率的に学ぶ

第1版　第1刷　2007年 9月25日	著　　者　國安　均
第23刷　2025年 2月10日	発 行 者　曽根　良介
	発 行 所　㈱化学同人

検印廃止

〒600-8074　京都市下京区仏光寺通柳馬場西入ル
編 集 部　TEL 075-352-3711　FAX 075-352-0371
企画販売部　TEL 075-352-3373　FAX 075-351-8301
　　　　　　　　　　振　替　01010-7-5702
e-mail　webmaster@kagakudojin.co.jp
URL　https://www.kagakudojin.co.jp

JCOPY〈出版者著作権管理機構委託出版物〉
本書の無断複写は著作権法上での例外を除き禁じられています．複写される場合は，そのつど事前に，出版者著作権管理機構（電話 03-5244-5088, FAX 03-5244-5089, e-mail: info@jcopy.or.jp）の許諾を得てください．

本書のコピー，スキャン，デジタル化などの無断複製は著作権法上での例外を除き禁じられています．本書を代行業者などの第三者に依頼してスキャンやデジタル化することは，たとえ個人や家庭内の利用でも著作権法違反です．

印刷　創栄図書印刷㈱
製本　藤原製本㈱

Printed in Japan　© Hitoshi Kuniyasu　2007　無断転載・複製を禁ず　　ISBN978-4-7598-1059-2
乱丁・落丁本は送料小社負担にてお取りかえいたします．

● 音声ファイルの内容 ●

トラックNo.	収録内容	本文ページ		トラックNo.	収録内容	本文ページ	
1	[ɑ]と[ɑː]の発音	3	C H A P T E R 1	42	化合物の名称(置換炭化水素)	29	C H A P T E R 2
2	[æ]の発音	3		43	（ハロゲン化物，有機リン化合物）	29	
3	[ʌ]の発音	4		44	（無機化合物）	30	
4	[ər]と[əːr]の発音	4		45	（アミノ酸）	30	
5	[ɑːr]の発音	4		46	人名反応など	31	
6	[ə]の発音	5		47	その他の反応名など	33	
7	[i]と[iː]の発音	5		48	学生・研究生活関連の単語	34	
8	[u]と[uː]の発音	5		49	学年・職位・学位などの単語	35	
9	[ɛ]の発音	6		50	教科名などの単語	35	
10	[ɔ]と[ɔː]の発音	6		51	実験関連の単語	36	
11	[ɔːr]の発音	6		52	Lesson1　構文 1〜10	38	C H A P T E R 3
12	[ai] [ei] [ɔi] [au] [ou]の発音	6		53	Lesson2　構文 11〜20	42	
13	[k]と[g]の発音	7		54	Lesson3　構文 21〜30	45	
14	[s]と[z]の発音	8		55	Lesson4　構文 31〜40	47	
15	[θ]と[ð]の発音	9		56	Lesson5　構文 41〜50	50	
16	[ʃ]と[ʒ]の発音	9		57	Lesson6　構文 51〜60	52	
17	[t]と[d]の発音	10		58	Lesson7　構文 61〜70	55	
18	[tʃ]と[dʒ]の発音	11		59	Lesson8　構文 71〜80	58	
19	[m]と[n]と[ŋ]の発音	12		60	Lesson9　構文 81〜90	60	
20	[f]と[v]の発音	13		61	Lesson10　構文 91〜100	63	
21	[h]の発音	14		62	Lesson11　構文 101〜110	66	
22	[p]と[b]の発音	14		63	Lesson12　構文 111〜120	69	
23	[j]と[w]の発音	15		64	Lesson13　構文 121〜130	72	
24	[l]と[r]の発音	16		65	Lesson14　構文 131〜140	74	
25	接頭語(成分比)	19	C H A P T E R 2	66	Lesson15　構文 141〜150	76	
26	（原子団などの数）	20		67〜70	Lesson 1　Atom	81	C H A P T E R 4
27	（その他）	20		71〜73	Lesson 2　Bond strength and lengths	85	
28	接尾語	21		74	Lesson 3　Hydrogen bonding	88	
29	乗　数	22		75〜78	Lesson 4　Catalysis	89	
30	元素の名称	22		79〜80	Lesson 5　Carbonyl compounds	93	
31	化合物の名称(アルカン)	24		81〜84	Lesson 6　Si face and re face	95	
32	（アルケン）	25		85	Lesson 7　Nucleophilic substitution	99	
33	（アルキン）	25		86〜87	Lesson 8　Schlenk equilibrium	100	
34	（アルコール）	26		88〜89	Lesson 9　Kinetics	102	
35	（アルデヒド）	26		90〜91	Lesson 10　Polymer	104	
36	（ケトン）	26		92〜93	Lesson 11　Supermolecular chemistry	106	
37	（カルボン酸）	27		94	Lesson 12　Photosynthesis	108	
38	（エステル）	27		95	Lesson 13　Cell -1-	110	
39	（ニトリルなど）	27		96	Lesson 14　Cell -2-	112	
40	（ヘテロ環化合物）	28		97〜98	Lesson 15　Bonding of alkenes to transition metals	114	
41	（炭化水素）	28					

スピーカー　Sukanda Tianniam／洪愛薇